P9-DHM-117

INGENIOUS KINGDOM

The Remarkable World of Plants

Stanley Wyatt

INGENIOUS KINGDOM

The Remarkable World of Plants

by HENRY and REBECCA NORTHEN

Line drawings by STANLEY WYATT

PRENTICE-HALL, Inc., Englewood Cliffs, N.J.

INGENIOUS KINGDOM
The Remarkable World of Plants
by Henry and Rebecca Northen

ISBN: 0-13-464859-5

Library of Congress Catalog Card Number: 76-110413

Printed in the United States of America *T*

Prentice-Hall International, Inc., London
Prentice-Hall of Australia, Pty. Ltd., Sydney
Prentice-Hall of Canada, Ltd., Toronto
Prentice-Hall of India Private Ltd., New Delhi
Prentice-Hall of Japan, Inc., Tokyo

This book is dedicated to all plant lovers
In the fond hope that through it they may

Wonder at the ingenuity of plants,
Revel in their beauty,
Delight in their "games,"
Smile at their "tricks,"
Marvel at their riches,
And be awed by their mysteries.

By the same authors:

Henry T. Northen:

Introductory Plant Science

Rebecca T. Northen:

Home Orchid Growing
Orchids As House Plants

Co-authored:

The Complete Book of Greenhouse Gardening
The Secret of the Green Thumb

Acknowledgments

We gratefully acknowledge the kindness of all those who provided such fine photographs for this book.

Contents

Introduction

"Botany" once meant little more than naming and classifying plants. Those activities are still part of botany—a rewarding part to the amateur nature lover, since we somehow seem to know a thing better when we can call it by name. But during the past two centuries, the scope of botany has increased enormously, first by the inclusions of plant geography, evolutionary development, and ecology until finally, as biochemistry probes deeper and deeper into the hidden mechanisms of the cell, botany has become a mere subdivision of the inclusive science of living things.

Of the two great categories of living things, the plants must have come first because only they can nourish themselves on inorganic materials and create the protein on which all animal life depends. In this sense also, "all flesh is grass." The amateur naturalist, though, usually knows a great deal more about the natural history of animals than he does about plants. He often thinks of them as just simpler and lower forms of life, interesting chiefly as the most attractive ornaments of the earth we live on. But science has discovered that what one is tempted to call the "behavior" of plants—their intricate, built-in responses to stimuli—is almost as complex as the animals'. In one respect, it is even more wonderful: plants have solved their problems and achieved their remarkable adaptations without (or at least we assume) any intelligence or consciousness. A true biologist would no doubt be horrified to hear us refer to their "ingenuity" or attribute "instincts" to them.

The present volume is, in fact, a great deal more than a simple introduction to the entire scope of plant science—not only to classification, evolution, and ecology, but also to what I have dared to call "plant behavior." Masterful in its rich, specific detail as well as in its generalizations, this book is non-technical only in the sense that it requires no previous knowledge to be understood. The reader can come away with a real grasp of what the science of botany—in the broadest sense—knows about the plant world.

Any prospective enthusiast of "plant behavior" might well sample Chapter 12, where he will find what's currently known about the chemical processes and built-in responses that account for such mysterious "timetables" as a flower's blooming season, and Chapter 8 for the fantastic, infinitely varied pollination devices which plants have developed during the last few hundred million years. (The earliest flowering plants trusted to the amazingly wasteful technique of merely scattering pollen to the wind, in the hope that some of it would by chance fall upon the waiting stigma of another flower of the species. Their descendents still do.)

It isn't strange that until the triumph of Darwinism, the so-called "Argument From Design" convinced nearly everyone that the universe had been planned and created by some all-powerful intelligence *outside* of nature. Indeed, the whole story as we tell it today is so extraordinary that a few persist in doubting—not that evolution took place, but that man is any closer to understanding the basic "hows" and "whys" than he was a hundred years ago. This book, then, is a marvelous record of where botanists have been, and where they still have to go.

Joseph Wood Krutch

INGENIOUS KINGDOM

The Remarkable World of Plants

chapter one

Ingenious Kingdom

AS you stand in a forest, you may hear your own heart beat, but you will hear no sound from the trees around you. No noise of motors or columns of smoke proclaim the processes going on within the trees as they raise tons of water from the earth to the top of their great structures and manufacture thousands of products within their own laboratories.

Man's body is far more complicated than a plant's, with its nervous system and various organs, and its marvelous coordination between mind and muscle. Man has even built machines to do much of his work. But for all of his inventiveness, he has not yet been able to manufacture the raw materials upon which his very life depends. The vegetable kingdom is the sole source of oxygen, sugar, cellulose, and other vital products; and mankind spends much of its time and energy harvesting these substances from plants and putting them to use.

Few shoppers stop to think that all of the food displayed on a supermarket's shelves has its origin in the plant kingdom. The fruits, vegetables, spices, cereals, coffee, tea, sugar, and flour come to us directly; dairy products, meats, fowl, and seafoods come from animals which rely on plants for food. Oddly enough, plants also form many substances which they do not seem to use themselves—but which man is finding to enrich his own life and ameliorate his ills. Scientists have not by any

Penicillin mold, *Penicillium chrysogenum,* a mutant form of which now produces almost all of the world's supply of penicillin. *(Charles Pfizer & Co., Inc.)*

means exhausted the numbers of these products. Drugstores offer not only the old plant remedies such as digitalis, morphine, belladonna, and colchicine, but the newer ones made by the lowliest of plants, the fungi. Until 1928, when Alexander Fleming noticed that no bacteria grew around the tiny islands of blue penicillium mold in his culture dishes, some people thought that research on these insignificant plants was largely fruitless. Now vast sums are spent each year exploring the possibilities of other fungi so that we may have more and ever newer antibiotics, and antiviral and antitumoral agents.

Coal, laid down by plants of ages past, gives us uncountable products for uses other than its contribution as fuel. Gas and oil are also products of ancient plants. The development of our modern industrial civilization was based on these three products, and nations still fight for control of their sources. Coal is now yielding chemicals which are in short supply from living plants, compounds from which come rubber, drugs, disinfectants, dyes, detergents, varnish, paint, explosives, and fertilizers. Some of the synthetic fibers such as dacron, nylon, and acrilon are made from chemicals derived from coal. Most of the plastics, from melamine to plexiglass, are made from coal tars combined in some instances with cotton or wood pulp and treated with various chemicals.

Any manufacturer would consider himself lucky if, as the plants do, he could obtain his raw materials at no cost. By photosynthesis, plants capture the energy of the sun to make sugar from carbon dioxide and water. From the simple sugars they make starches, cellulose, fats, oils, and waxes. By adding minerals from the medium in which they grow, whether soil or water, they manufacture proteins, enzymes, pigments, resins, and vitamins. Man has learned to cultivate those plants that are particularly rich in carbohydrates, fats, minerals, proteins, and vitamins, and he selects the parts that are most palatable—seeds, fruits, roots, stems, leaves, or buds.

But for all of the expanse of farms, orchards, and grazing lands, human beings really use but a tiny fraction of the food made and stored by the world's plants. Naturally enough, plants use most of it themselves. It has been estimated that they make 375 billion tons a year! Of this amount, 90 percent is produced by aquatic plants. (We benefit from some of this as seafood, of course, but this great source is as yet hardly touched. Scientists are trying to find ways to use more of the products of the sea to supplement our own diet and that of domestic fowl and livestock.) Of this huge total, including crops planted by man, all animal life uses about 2 percent. The human race uses about two-tenths of one percent, partly because most of it is unsuitable for our consumption. Our systems are equipped to handle only small parts of a tree (its nuts or fruits, for instance) or of grass (grain).

There is a general idea that since animals are mobile, they are more active physiologically than plants. Actually, however, activity is relative. Consider a woman of 120 pounds who enjoys gardening. She may feel a touch of envy for the plants, which can just sit there absorbing the sunshine and growing without seeming to do any "work." But one way to measure activity is by seeing how many calories an organism spends each day. For the 120-pound housewife an expenditure of 1,800 to 2,400 calories is about normal, but 120 pounds of carrots expend some 3,200 calories a day. Pound for pound, it is clear that they do more "work" than she does.

In carrots and in the human body, the process of respiration is practically identical. About fifty chemical steps are necessary to obtain energy from sugar, the basic foodstuff for all living things. Some of these steps are carried out in minute granules within living cells, called mitochondria. The final product is a chemical called ATP (adenosine triphosphate), which acts like a minuscule storage battery. From this little battery, energy is released as needed for all of the activities in living organisms—for growth, for synthesis of chemicals, for absorption and movement of materials, and, in animals, for muscular and nervous reactions.

All living things must have some way of obtaining nutrients and distributing them to their tissues. Living cells can operate only in a fluid medium, and this is true of both plants and animals. The simplest system for plants is that of the floating algae, which have only to absorb nutrients and oxygen from the water, salt or fresh, that bathes their cell membranes. When cells were incorporated into the many-celled organisms that constituted land plants, there had to evolve a system of roots to penetrate the soil and obtain the necessary water and minerals. No one has ever counted the roots on a tree, but a study was made of those of a single rye plant. It was figured that their total number was 13,815,762, and their combined length 387 miles. From the tips of the finest roots, tiny projecting root hairs one cell thick push their way between soil particles and do the actual job of absorbing water and minerals. The total number of root hairs on this rye plant was estimated at 14 *billion,* with a total length of 6,600 miles.

Minerals and water are transported to all the cells of a plant through a system of end-to-end tubules that extend from the roots to the tips of every branch and leaf. At the same time, sugar made by the leaves travels through another parallel set of tubules reaching every cell down to the most remote root hair. Thus there is a two-way system carrying the life-giving fluids and chemicals.

Just as animals fatten themselves in preparation for winter, so also do plants plan for the seasons and for the perpetuation of the species. During their best "earning" months—the summer—apples and lilacs, willows and oaks, tulips and daffodils work at top speed to produce extra food and store it in roots, stems, and bulbs for growth the following spring. During the summer they also develop the next year's leaves and flowers, enclosing them in tight buds that enter a resting state during which they will not be injured by winter's cold. Their profound sleep cannot be broken by the continued warm days of fall, nor by the mere passage of time. They must first experience winter. Internal changes go on in the resting buds without which they cannot respond to the wakening touch of spring. When the changes are completed, the masses of crocuses and jonquils, the oceans of blossoms in orchards, the pussy willows along the streams, and the fresh young leaves reveal the planning of the previous summer and the benefit of winter's discipline.

The work and planning that go into seeds, and the devices plants have invented for their dispersal, are perhaps one of the most marvelous aspects of plant life. Many plants achieve their pollination by

The orchid, *Trichoceros parviflorus,* imitates the female of a species of fly so exactly that the males attempt to mate with it.
(*Rebecca T. Northen*)

some of the neatest tricks ever conceived by living things—they lure insects, birds, and animals to do their work for them, sometimes even imitating the insects themselves or acting as "decoys" for members of the desired species. This is all the more remarkable because the plants as a group evolved *before* animals came into being. Obviously, many plants evolved features that enabled them to make use of the newly arrived potential servants.

Plants are all around us, no matter where we live, and they have become adapted to grow where man can barely survive. In the humid tropics they crowd upon each other, some climbing and dwelling on others and forming aerial gardens. In the Arctic and in the austere alpine regions they huddle in dense mats or occupy rock crevices where they give brilliant displays of color in the brief summer. They are able to live in the deserts—which are by no means the barren wastes we sometimes picture—and here they have evolved weird and fantastic shapes and habits. They carpet the plains and prairies with grasses and clothe the mountains with trees. They live in lakes and streams, and in the oceans to a depth of 600 feet. Some that live in water are flowering plants, others are algae, ranging from microscopic size to forms many feet in length.

Most of the plants we know depend on chlorophyll, but there are nongreen ones as well—the fungi and bacteria. We can see some of them—the mushrooms, the shelf-fungi on rotting logs, the molds that cover stale bread and overripe fruit—and we know others as the source of antibiotics. Many we cannot see because they are microscopic in size, but we are familiar with bacteria as agents of disease. Other fungi, the yeasts, ferment wines; still others give flavor to cheese. Those that live in the soil do double service: not only do they clean up dead plant and animal material, but by digesting it they return essential minerals to the soil, keeping it fertile and protecting its structure. Microscopic algae live along with the fungi. In just one pinch of soil there are millions of bacteria, fungi, and algae; in one drop of pond water there may be thirty or forty different species.

As you travel across the country, you see one type of vegetation giving way to another. The panorama slowly changes as climate and terrain work their influences. The landscape has much to tell you, if you but read it as you go. As you pass through the forested East, large tracts of woods softly covering mountainsides and valleys remind you of what once was an unbroken expanse of trees; across the great grasslands of mid-United States the prairies blend into the plains; over the Rocky Mountains tundra comes close to snowy peaks, and across the vast deserts of the West the land is gashed and molded by wind and flash floods and vegetation is sparse.

The type of native plant community reveals the nature of the soil and the amount of precipitation, indicates the kind of animal life that accompanies the land, and suggests its potential uses. In each habitat the beautiful and the bizarre, the gentle and the fierce, the microscopic and the gigantic interact so that some of each kind can continue in perpetuity. A subtle and crucial web exists between plants and animals and among the plants themselves—a web that functioned very well before man became a part of it. Hindsight makes us realize how often man's alterations were unwise; the land itself could have warned those understanding enough to see. With experience as our guide, the future is more promising.

In a stable plant-animal community the various inhabitants maintain a long-range balance, usually recovering from temporary upsets that naturally occur from time to time. No organism can live alone. Even a plant in a pot in your living room is the center of a colony; the soil is teeming with organisms that cooperate with it and with each other. Every community has its characteristic inhabitants: plants that are adapted to the conditions of the region and animals in turn that are adapted to live with them.

Human beings seem to feel most comfortable when they have plants around them. A man usually desires a piece of ground to call his own, where grass and flowers reflect both his handiwork and nature's. He decorates his cities with parks and the foyers of his office buildings with potted plants to soften the effect of steel and concrete. Plants give a touch of serenity to busy places, partly because of their color and grace, but isn't it possible that they subconsciously remind us of our bond with nature, reminding us that frantic, hectic human affairs are not all that matter on this earth?

chapter two

The Coming of Life

IT was life itself that tamed this planet. Without the influence of living things it would still be but barren rock. No soil would soften its contours; lifeless waters would lap on sterile shores. But it was the earth that first invited life to develop. Its surface gentled down to a temperature in which living things could exist. It offered water and life-building materials. It provided a protective atmosphere. It orbited a congenial sun at just the right distance to benefit from its light and warmth. Yet how and when did life begin? What events came about to quicken nonliving matter, to bring about the first primordial organisms? And how did they lead to the varied forms that dwell on earth today?

We do not know when life first actually appeared, but we do have clues as to how it may have begun. From the fossil record we know that a wealth of life already existed three billion years ago. According to a prevailing theory, our solar system was formed five billion years or more ago by the condensation of interstellar matter spinning in space. A central mass became the sun and peripheral masses the planets. The new-born earth was tortured by fierce temperatures, raging volcanic activity, and cataclysmic storms. It has not yet calmed down completely, but for some time longer than those three billion years it has been habitable.

Theoretically, the primeval atmosphere was rich in hydrogen, ammonia, water, and the carbon compound methane (CH_4)—also called marsh gas. Reacting to energy from the sun and from lightning and radioactivity, the molecules in the atmosphere formed an array of chemicals that became dissolved in the oceans. Thus the waters of the earth were enriched with sugars of various kinds, amino acids, lipids, and nucleotides, all of which were relatively small molecules. That they could have been formed is substantiated by modern experiments that have duplicated them in the laboratory under conditions simulating those that must have prevailed during earth's early history. In one such experiment, when an "atmosphere" of ammonia, water, hydrogen, and methane was enclosed in an evacuated tube and exposed to electric discharges simulating lightning, amino acids and proteins were formed.

The compounds accumulated in the ocean because there were no organisms to use them and no oxygen to break them down. (If such compounds were formed today by nonbiological methods, they would be quickly consumed.) Over millions of years the oceans became an increasingly rich solution of chemicals. Small molecules joined together to form larger ones; we say they *polymerized*. Groups of sugar molecules joined together to form starch. Amino acids linked up to form proteins. And then came the spark of life—nucleotides joined together to form nucleic acids, the stuff of which DNA, the programmer and director of all of life's processes, is made.

The appearance of DNA (deoxyribonucleic acid) was full of portent, for it supplied what had been lacking in the chaos of the primordial seas—the ability to reproduce in an orderly manner and to control by means of a built-in code the formation of specific proteins and enzymes. The enzymes programmed by DNA were in their turn capable of catalyzing specific chemical reactions.

Marvelously, DNA is constructed of just four kinds of molecules, four chemical building blocks, hundreds of each lined up to make a long strand. These four chemicals, selected from among the millions present in the waters, have proved to be the only ones capable of handling the genetic code; they remain unchanged today. Bridge players know that a tremendous number of combinations can be achieved from just thirteen units of four different suits. The possible combinations for DNA are astronomical. The sequence in which its molecules are arranged forms the code by which it programs a particular activity.

DNA is actually a double strand in the form of a spiral ladder, and in the double strand the four kinds of molecules are always paired in a certain way opposite each other. When DNA reproduces itself, it zips open up the middle, breaking the bonds in the center of the rungs. Each half then acts as a template for the formation of another partner strand identical to the original one. Millions of strands of DNA must have been formed in the primordial ocean, each with its own code.

Even though the stage was now set for the appearance of the first primitive one-celled organisms, it was undoubtedly millions of years before a fully functional one was formed. We have to imagine a membrane forming around and enclosing strands of DNA, proteins, minerals, water, and other substances present in the ocean. There had to come about a happy combination of strands of DNA capable of working together to coordinate the activities of a primitive cell in order for it to maintain itself as an organism. Once it was brought about, however, its DNA would have directed its functions and reproduction. The organism would then have been able to absorb ready-made nutrients from the water, incorporate them into its protoplasm, and grow. When it reached a size that was no longer efficient, the DNA would have directed its division. Its contents, including duplicate strands of DNA, would have been apportioned equally to the daughter cells. On and on, through division after division, it would have multiplied, the DNA carrying its traits to each new generation.

Although DNA reproduces itself with remarkable precision and has been doing so ever since the first strands were formed (we may have some very ancient DNA in our own cells), changes came about from time to time. A few extra pairs of the chemical building blocks would be added to a strand, or the position of a pair would be reversed, changing its code. Even a small change could cause it to program a slightly different characteristic in the organism. The change might often have brought about a flaw that would cause the organism to fail. But if it were a beneficial change, especially one that gave it some advantage, the organism with the new characteristic would thrive and reproduce itself. Thus the primordial cells gave rise to new types which, through trillions of mutations and millions of years, arrived at the variety of living forms on earth today.

Mutations are still occurring, but since we live in such a brief moment of time we see very few. One that we are familiar with was the spontaneous appearance of a peach without fuzz, the nectarine. Mutations have occurred in insects by which they have become resistant to insecticides. Strains of bacteria have become resistant to antibiotics.

In the meantime, the first cells developed a means of respiration, a method of releasing energy from some of the foods they absorbed—from sugars for example. But this respiration would have had to be anaerobic, for there was no free oxygen in the early atmosphere. Some organisms today still carry on respiration in this primitive manner, releasing energy from food in the absence of oxygen. The yeasts use sugar and produce alcohol and carbon dioxide, and some bacteria also function without oxygen. Anaerobic respiration added a new gas to the atmosphere and to the water: carbon dioxide.

Carbon dioxide set the stage for a process that was to change the aspect of the earth, photosynthesis. The ready-made materials in the ocean were becoming exhausted by the rapidly increasing populations of organisms. Fortuitous mutations in some organisms gave them pigments that could absorb energy from the sun and enabled them to synthesize their own sugars from carbon dioxide and water. The cells containing these pigments were the first green plants, primitive algae, and the pigments were early forms of chlorophyll. Starting with the sugars they could now make, and adding minerals from the water, the algae were able to synthesize all of the chemicals necessary to life. Living things no longer had to depend for food on molecules formed in the past; they could create their own.

From this time on, all organisms would be supported directly by photosynthesis. In a sense, life would operate on solar energy, a bounteous source indeed. The sun emits enormous amounts of energy, resulting from the fusion of hydrogen into helium, 564 million tons each second. We need not fear that the supply will become exhausted any time soon. It is estimated that there is enough for several billion more years. If the earth evolved five billion years ago, and has another five billion to go, we are now at about the halfway point in its existence.

But the algae provided more than food. They furnished oxygen to the atmosphere and to the waters, enabling aerobic respiration to evolve. Ever since, plants have been the only source of free oxygen. They used the oxygen they released for their own respiration and at the same time made enough to allow the evolution of animal life. Respiration using oxygen releases far more energy from food than does anaerobic metabolism. Now we are endangering the supply by burning fuel in cars, homes, and industries at such a high rate that photosynthesis may not be able to keep pace. The amount of oxygen we tie up may exceed the amount produced. Surely we see the necessity to guard the plants of the world.

All of the essentials for continued life had now evolved. The molecules formed in the waters had lined up to form the first materials of life, DNA had taken over their control, anaerobic respiration had made possible and sustained the primitive organisms, and photosynthesis had revolutionized the life process.

The primitive cells lacked internal specialization, but specialized structures evolved as time went on. The DNA became located in a nucleus, a spherical body separated from the other cell contents by a membrane. Strands of DNA became incorporated with protein to form genes, and these in turn grouped together to form chromosomes, which gave a more efficient mechanism for reproduction and the parceling out of hereditary traits to offspring. Each strand of DNA would now operate by manufacturing a messenger, RNA (ribonucleic acid), to move out from the nucleus into the cytoplasm and carry out the activity dictated by the DNA.

In the cytoplasm many specialized structures developed which made of the cell a most remarkable little self-sustaining engine. The light-absorbing pigments, chlorophyll, which were at first diffused throughout the cell, became localized in minute organs called chloroplasts, which carried on photosynthesis more efficiently. Centers for respiration, called mitochondria, became the powerhouses of the cell. The synthesis of proteins and enzymes became localized in granules known as ribosomes, the centers where RNA carries out the programs set up by DNA.

It was only a small step from the first one-celled plants to the first one-celled animals. All that were needed were a few mutations to enable some to use others as food. Most of the genes would have remained unchanged because many of the life processes are the same for both plants and animals. Even in the complex plants and animals of today the fundamental processes are the same and the genes to control them carry the same codes. Human beings are thus more like plants than you might think, and in fact, many of our genes are identical to those of the plants growing in our gardens and the organisms within the soil. Whether an organism turns out to be a bird, a man, a mushroom, or a tree depends only on the arrangement of the building blocks of its DNA.

The progression from one-celled organisms to more complex ones began with a mere clumping together of similar cells to form colonies, in which each member continued to carry on its independent activities.

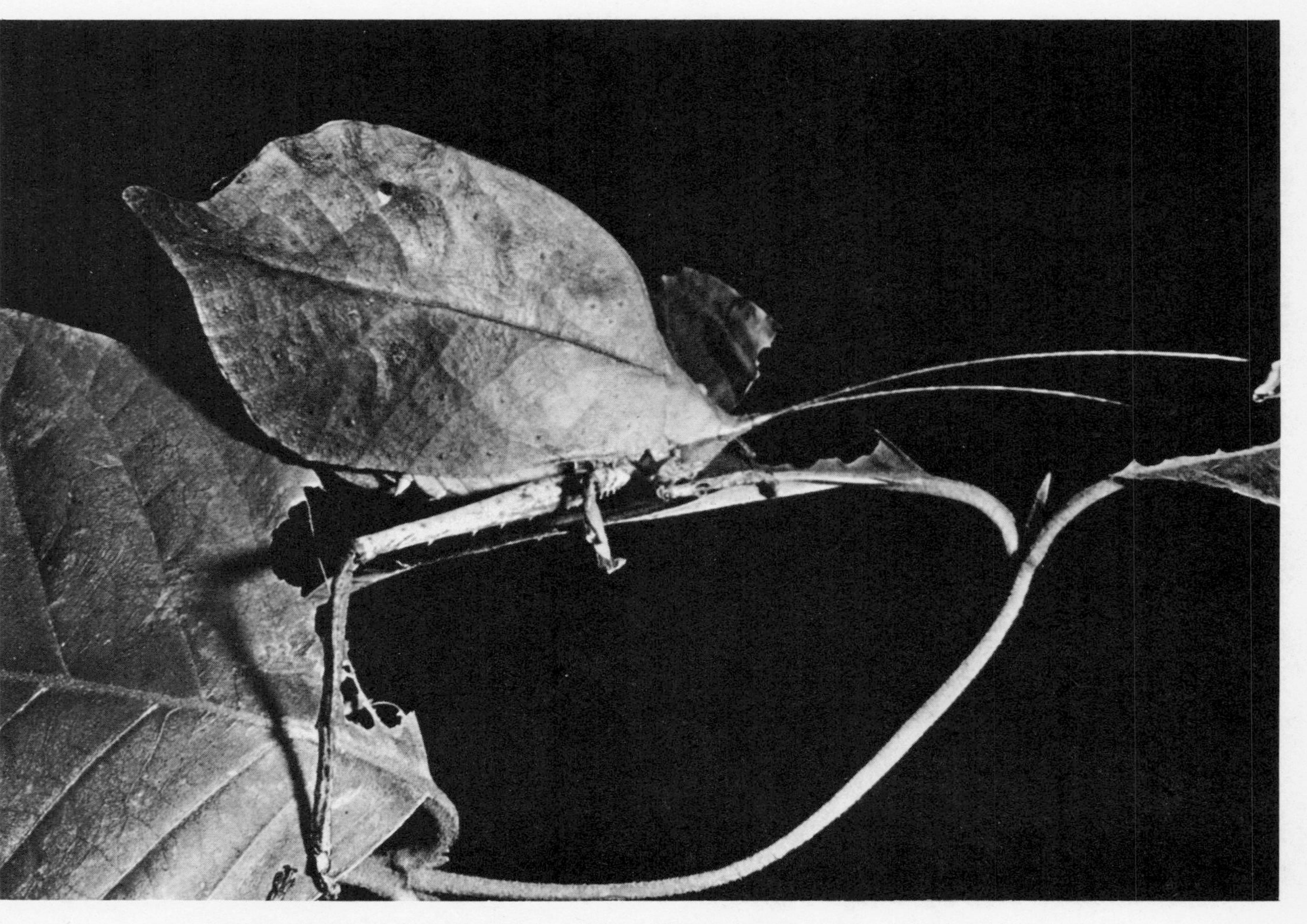

Perfect mimicry is shown by this grasshopper that looks like a leaf. (*Gerald Lang*)

From some of these arose organisms in which groups of cells became organized into specialized tissues. Mutations along the way led to many "experiments," some of which brought about quick extinction. Others led to organisms that have continued unchanged to our time, in which no further evolution took place. Some went through subsequent mutations leading to long evolutionary lines. Throughout all time, the tissues of life, which seem so fragile, have proved more durable than the hardest stone. Oceans have come and gone, mountains have been formed and eroded away, but life has persisted and become more magnificent and more varied.

The capacity of living things to adapt to new environments and changed conditions is one of the marvels of nature. The wonder is not only that so many forms have evolved, but also that the forms are so neatly suited to all of the available niches earth offers.

We are now on the threshold of a new epoch. Man has moved into space and is eager to explore other planets. Will he encounter life, and if so in what forms? In order to anticipate where life may occur elsewhere, he might use some clues derived from experience on earth.

The absolute requirements for life as we know it are temperatures in a certain range, a supply of water, a suitable atmosphere, certain minerals, and a source of carbon. The lowest possible temperature known to man is 459° F. below zero, at which all chemical activity would cease. The highest is 5,400,000,032° F., at which matter itself would cease to exist. Life on earth can survive only in a narrow band between these extremes—from 459° below zero to 302° F. above; and life forms are active in an even more narrow band. Living things survive below-freezing temperatures in a dormant state; they are not active when their cell contents are below freezing. Scientists have subjected spores to 454° F. below freezing, and some have survived. Some spores may survive temperatures as high as 302° F. but neither are they active, nor is any life activity possible, at that extreme.

The so-called biokinetic range (the range for activity) is from about freezing to about 122° F. Exceptional organisms can grow outside this range, among them certain algae and bacteria in hot springs where the temperature is close to 185° F. and the water is steaming; but they are few. Temperatures on the earth's surface have not been known to drop below 159.8° F. below zero, nor, except near volcanoes and in hot springs, to exceed 134° F. above.

Temperature alone would exclude life from all the planets of our sun except Earth and Mars. Jupiter, Saturn, Uranus, and Neptune are far too cold, and Mercury is too hot on one side and too cold on the other. Venus, once a candidate in our imagination, has been proved to have a temperature of 932° F.—impossibly hot.

That leaves only Mars as a potential host for living things. Its temperatures are compatible, ranging from —212° to 86° F. Photographs and other data taken by Mariner IV, VI, and VII reveal that Mars has a moonlike landscape speckled with more than 10,000 craters, that it lacks a magnetic field, and that its atmosphere is very thin. There is as yet no evidence that during its long history (about five billion years), it has ever had sufficient water to form streams, to bring about erosion, or to fill oceans. Hence Mars seems to be an inhospitable place for life (at least beyond some very primitive forms).

It appears, therefore, that Earth is the only true bioplanet of our solar system. But what of planets of other suns? Surely some of those other stars have planets, perhaps some with environments compatible with life. Biochemists believe that if conditions elsewhere had been in the past similar to those on ancient Earth, life would have evolved.

At least 100 quintillion—100,000,000,000,000,000,000—stars can be seen with telescopes, and there are countless ones that can't yet be viewed. Not all stars have planets, perhaps only one in a thousand, which gives 100 quadrillion—100,000,000,000,000,000—possibilities. Of these, one in a thousand or 100 trillion—100,000,000,000,000 —might have temperatures compatible with life. Considering that an atmosphere is also necessary, and again that one in a thousand might supply it, that narrows the possibilities down to 100 billion—100,000,000,000—that might provide both proper temperatures and an atmosphere. Even given these conditions, life might not have evolved, but say it did come about on one in a thousand, that would mean 100 million—100,000,000. These are the speculative figures of a prominent astronomer, but if they should come close to actuality, how can one imagine the variety of life forms that might have come about through 100 million different processes? Some would still be quite primitive while others would have advanced far beyond ourselves. If the one-in-a-thousand ratio were projected further, it would mean that there might be 100 thousand that had reached a stage of civilization perhaps at least the equal of our own. Confident that this is so, some scientists propose building huge radios to receive signals from outer space. They believe that man will one day communicate with intelligent creatures in other parts of the universe.

Still, whatever stages of evolution may have been reached elsewhere, it is interesting to ponder how any world could have evolved higher forms of life without the help of the most primitive of one-celled plants.

chapter three

Algae

ALGAE, the most ancient of the earth's flora, have witnessed the coming of all other plants and of all animals. They sprang from the evolving primordial cells and have been found as fossils in rocks three billion years old. As the algae evolved, many types arose that we see in the water and soil today. The blue-green algae are so primitive in form that they must still be very similar to those primordial cells. One evolutionary line led from the early algae to the mosses. Still another led to the ferns and, through them, to all the higher plants of our world. Yet a different line took off from the primordial cells to form the bacteria and fungi, and these proliferated among themselves without giving rise to other organisms. As a result, some of the bacteria have much in common with the blue-green algae.

There are now 20,000 species of algae, and the variety of forms is extraordinary. From single cells that live alone or form long chains or filaments, they range up to the massive and complicated seaweeds of the world's oceans.

All algae have the green pigment chlorophyll, but it is often masked by pigments that color the entire plant violet, blue, blue-green, yellow, red, or brown. Some groups of algae are known by the predominant color—the blue-green and the green algae, for instance. But even within the same group there are forms in which a different color predominates. The Red Sea owes its color to a blue-green alga that is red. Some of the formations in Yellowstone Park derive their red hue from blue-green algae. The travertine terraces of Mammoth Hot Springs are colored yellow, pink, and gold by other blue-green algae (the travertine itself is built up by calcium they deposit). Snow that lies a long time at high elevations of the Rocky Mountains is colored watermelon-red by a green alga with a red pigment, *Chlamydomonas nivalis* (and curiously, it even smells like watermelon), while snow in Yellowstone Park is colored green by one that really *is* green, *Chlamydomonas yellowstonensis*. A brown pigment predominates in most of the ocean algae, but some are red, violet, or purple.

The blue-green algae of today are all unicellular and have no nucleus. The various cell components, including chlorophyll, are dispersed in the cytoplasm. Other algae are a bit more advanced. They are likewise unicellular, but they have a nucleus. Their chlorophyll is contained in chloroplasts, and they have special granules to carry out respiration and protein synthesis. Still more complicated are those algae whose cells are held together in colonies, and most advanced are the multicellular ones such as kelp, in which cells have specialized functions—some ("holdfasts") to anchor the plant, others to form a stalk, and still others to form expanded leaflike structures.

Algae of today demonstrate all of the stages in the development of sexual reproduction. The simplest reproduce merely by dividing in two. Some periodically go through what is called conjugation, in which cells come together and exchange genetic material. Afterward they separate and continue to reproduce by splitting in two. The dawn of true sexual reproduction occurs in species a bit more advanced, which produce cells that fuse together to form a new individual. But still the joining cells are of the same size and shape; they are not differentiated into sperms and eggs, nor are there specialized structures for their formation.

Fully differentiated sperm and egg cells and the structures to form them appear in still further advanced kinds. The sperms (some mobile and some not) are tiny, and great numbers can be made. Those that go to waste—that is, those that do not fertilize eggs—do not represent a great loss of materials. The egg cells, however, are large and store food for the first stages of embryonic development. They are produced in fewer numbers, since the large number of sperm cells assures fertilization.

Many fascinating hours can be spent examining a drop of pond water under a microscope. In even this tiny volume there can be dozens of species of algae and of microscopic animals that feed upon them. You may have difficulty telling the plants from the animals. Nature has been just as lavish in designing this microscopic world as she has the one that is commonly before our eyes. In the drop will be diatoms, a kind of alga so beautiful that they merit the name "jewels of the waters." Each is a one-celled plant that lives in an exquisite little glass house made of silica. The house is really a box whose bottom fits up inside the top, and both surfaces are decorated with radiating lines and dots. Some diatoms are triangular, some round, some fan-shaped, and others are shaped like ships or crescents. So fine are their markings that they are used to test microscope lenses. They move through the water by cytoplasmic streaming; the cytoplasm is exposed to the water through slits and moves the diatoms as if by miniature tractor treads.

Diatoms are among the most important food for microscopic animals in both salt and fresh water, and therefore significant in the food chain that leads to higher animals. Great deposits of diatom shells laid down in the ancient seas and now condensed into fossil-bearing

Diatoms, "jewels of the waters," reveal under a microscope their beautiful shapes and the extremely delicate markings of their silica shells. (*Smithsonian Institution*)

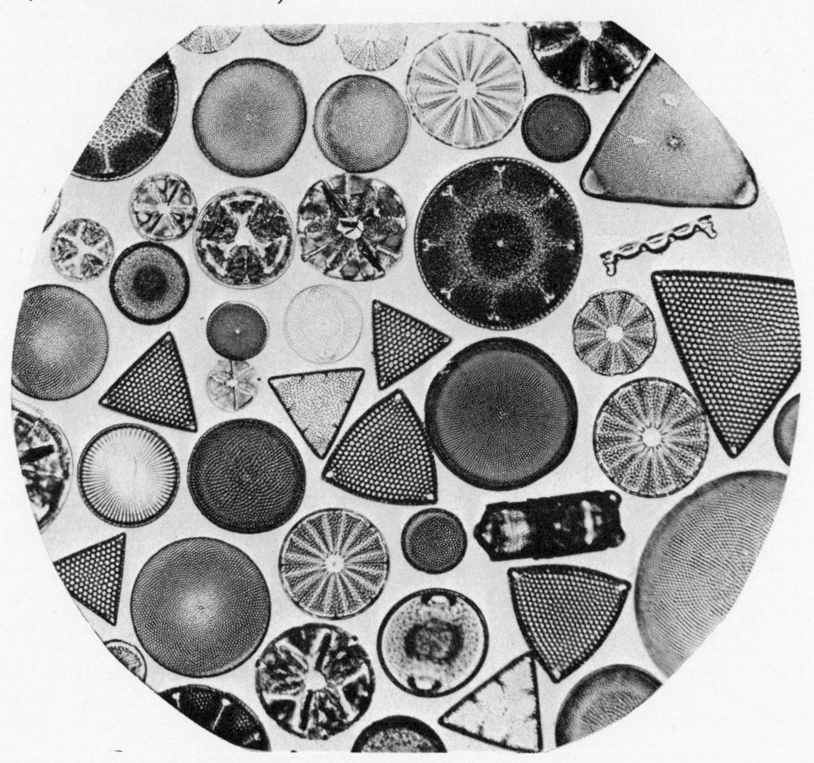

rocks such as those at Lompoc, California, attest to their abundance. When a diatom divides, one new cell receives the bottom of the box, the other the lid. Each then makes a new bottom. They thus become smaller with each generation until they go through the process of conjugation, which seems to give them renewed vigor, after which they return to their former size and begin again.

As you watch through the microscope you may be startled to see a large green ball roll rapidly across the field of vision. That would be *Volvox,* a colony of cells held together in a sphere, each cell taking care of its own functions. Many filamentous algae will also be present: cylindrical cells joined end to end in a chain. One will probably be *Spirogyra,* which is recognized by a spiral ribbon-shaped chloroplast in each cell. You can also see *Spirogyra* floating in mats on ponds and streams, its filaments as fine and silky as hair. Other filamentous algae have different arrangements of the chloroplasts, and some feel rough or slippery to the touch.

Darting, swimming, or whirling through the drop will be many delightful forms that move by flagella, little taillike or oarlike structures. Such are *Chlamydomonas* and *Euglena,* elongated pear-shaped forms with a light-sensitive spot that looks like a red eye, and the quite similar dinoflagellates. Some of the ocean-dwelling dinoflagellates cause the "red tides" that sometimes poison fish. One of them, *Noctiluca* (literally, "night light") produces the luminescence of the sea. The dinoflagellates might be classed as animals except for the presence of chlorophyll, which joins them to the plant kingdom.

The tiny animal forms that busily feed on the algae are equally fascinating and charming. In lakes and streams algae provide food for larger animals as well as for microscopic ones, among them snails, small crustaceans such as fairy shrimp, and the larvae of many insects —caddis flies, mosquitoes, and mayflies. These and other small animals are food for small carnivores such as diving beetles, water scorpions, backswimmers, damsel fly nymphs, and dragon fly nymphs. Both the small herbivores and carnivores are eaten in turn by fish. Meanwhile, the algae and other water plants keep the water well oxygenated. You can know that fishing will be good where clear water flows over algae-covered rocks. In the oceans, as well, algae provide oxygen and food for the great variety of animal life.

Ocean shores offer never-ending fascination. We may enjoy strolling along a sandy beach when the tide is out, collecting shells and seaweeds left stranded upon it, watching the little crabs that venture out of the water to capture smaller creatures, and the shore birds that catch and eat both. Little boys like to pick up a long ropelike seaweed with a ball on its end and use it like a whip. This is *Nereocystis,* washed in from its mooring in deeper water. In life the ball is held upright in the water on the slender stem, and from it spread leaflike fronds which are torn away by the surf. Sand does not offer a stable anchorage for algae, however, so in order to see them in all their magnificence we must explore a rocky shore. Here will be a kingdom of great beauty with seaweeds of green, brown, purple, red, and lavender. A single rock may support a dozen kinds. Perched where it is covered and uncovered by the tides twice daily is *Postelsia,* which looks like a little palm tree and has a rubbery trunk that bends double when beaten by the surf. *Fucus,* which has long, slippery branched ribbons and little air bladders, hangs from the rocks. Out beyond the low-tide mark or in pools are kinds that remain submerged—*Polysiphonia,* with its red or purple fernlike fronds, and *Chondrus,* a violet endive-shaped form called Irish moss.

In the intertidal zone, the strange world that lies between the limits of high and low tide, many animals live on and among the algae. Where they are exposed to the force of the surf they cling so tightly to the rocks that they can scarcely be pried loose—limpets that look like peaked Chinese hats, chitons that resemble huge legless pill-bugs, mussels, barnacles, starfish, and sea anemones. When the tide is out they close their shells or pull in their appendages or cling immobile until water covers them again. A rock that seems to be covered with nothing but a soft jelly will, when the tide comes in, burst into bloom as the sea anemones, whose soft bodies made the "jelly," spread their petallike arms; they are animals but look like plants. In depressions among the rocks the receding water leaves pools where the intertidal life can be seen in all its beauty. In the crystal-clear water are barnacles moving their feathery appendages in quick grasping motions to catch any small creatures swimming by, starfish slowly gliding along on tube feet, crabs lurking under the edges of rocks or peering out from under seaweeds, sea anemones looking like pink or yellow or white asters, spiny sea urchins half buried in the holes they scrape out for themselves, and hermit crabs dragging their houses behind them, houses made of snail shells they have found or have stolen from each other. And all of these dwell in a garden of feathery or fernlike algae.

Farther out in the ocean where the water is deeper lives the giant kelp, *Macrocystis*. It grows to be 150 feet long, anchored on the bottom

A tide pool is a garden of algae of myriad forms and colors, some draped from the rocky walls and some growing upright from its floors. Among the plants live marine animals, both free-moving and sessile.
(*American Museum of Natural History*)

and spreading its attractive fronds on the surface. *Macrocystis* is harvested on a large scale for the algin it contains, a substance used as a gelling, stabilizing, and sizing agent in the paper, dairy, pharmaceutical, and candy industries.

While most seaweeds grow anchored to some support, one alga in particular just floats on the surface. This is *Sargassum,* whose long-branching fronds float in the Sargasso Sea, a 2,000,000-square-mile area in the Atlantic Ocean between the Bahamas and the Azores. The island of Bermuda lies in the middle of it. The Sargasso Sea is a curious sea within the ocean, a body of water that rotates like a giant eddy, dragged round and round by a circular system of ocean currents —the Gulf Stream and North Atlantic current moving on its west and north sides, the Canaries current on its east, and the Equatorial drift on its south. It is composed of warm water to a depth of about 3,000 feet, riding on the top of the cold water below. The floating *Sargassum* is never as dense as ancient legends pictured it and does not offer the dangers mariners long feared. Legends have a way of building upon worries, and the fact that mariners of bygone years found the Sargasso Sea different from other parts of the ocean may have fostered their exaggerations. For all of its wealth of seaweed, the Sargasso Sea is a desert. Perhaps the lack of the usual fish and such creatures as porpoises made the early navigators conscious of its strange emptiness. There are a few creatures that live with the *Sargassum*—some shrimps, crabs, molluscs, and a few small fish—and they feed on the algae and on each other. There are microscopic algae in the upper levels of the water, but all of these are in far fewer numbers than in the rest of the ocean. As one biologist put it, the Sargasso Sea is the clearest, purest, and biologically poorest of ocean waters.

Elsewhere, in the open ocean, most of the algae are microscopic one-celled forms, primarily diatoms. Each quart of sea water contains a million or more, and for every square mile of ocean surface they produce annually about 2,410 tons of food (compared to 920 tons per square mile of average farmland, and 1,500 tons in an average forest). The greater productivity of the ocean results from its more uniform conditions, relatively constant temperature, and absolute lack of drought. The oceans are a tremendous storehouse which has yet to be used to its full capacity. In recent years there has been a surge of interest in marine biology, and the ocean's great potential is coming to be appreciated.

The algae in the ocean are confined to the lighted upper zone, which extends down from the surface to 300 feet, and to 600 feet in very clear water. The numbers of algae gradually decline, however, as the light diminishes at these depths. Squids, whales, and fish such as the tarpon, tuna, and salmon are animals associated with this lighted zone. No green plants grow below 600 feet, but many animals live in the mysterious domain of perpetual darkness, among them the octopus and, of course, fish that possess light-emitting organs that enable them to see their prey. Some of these feed on creatures that drift down from above, and others move upward during the night to feed on organisms in the upper zone.

Algae can sometimes be a nuisance. They multiply in any water upon which light shines—birdbaths, aquariums, swimming pools, stock watering tanks, the always-damp glass of greenhouses, and so forth. When they attain excessive numbers in reservoirs, they may impart an unpleasant taste to the water. Bathing beaches can be made unattractive by masses of algae, which may discolor the water as well as bother the swimmers. In lakes and ponds polluted by sewage or enriched by agricultural fertilizers, their population explodes. This is a first step in a sequence leading to a very bad situation. As the algae increase, so do the oxygen-respiring bacteria that feed on their remains. After a while the bacteria use up all the available oxygen, and both they and the animal inhabitants of the water die. Then the anaerobic bacteria, kinds that can live without oxygen, begin to multiply, giving off foul odors that make a cesspool of a once pleasant lake.

Poisonous tides in the ocean are a recurring catastrophe of nature. They can occur anywhere in the world, and have been known for thousands of years. Perhaps the most famous in ancient times was described in the seventh chapter of Exodus: "And all the waters of the Nile that were in the river were turned to blood, and the fish that was in the river died; and the river stank, and the Egyptians could not drink of the river. . . ." Nearly the same words have been used in our day to describe mysterious instances of dead and dying fish piled in heaps along beaches, accompanied by a peculiar color of the water, sometimes yellow, often red. The cause of the poisonous tides has been found to be one or two species of dinoflagellates. Most of the many kinds are not poisonous, but when there is a population explosion of those that are, all of the fish in the immediate area are killed off. Fish that do not get enough of the poison to be killed themselves may be poisonous for human consumption. The poison is similar to that which causes botulism.

The reasons for the population explosions are not always clear. One clue, however, is that these small half-plant, half-animal organisms are extremely sensitive to small amounts of certain chemicals. Another is that they reproduce more rapidly in slightly diluted sea water. The explosions may be caused by runoff of water into the ocean following unusually heavy rains on land, which would dilute the sea water and also wash more phosphates into the ocean, giving them extra nutrients. Upwelling of currents in the open ocean bring up differing amounts of chemicals to the surface. Whatever the cause, the result is to bring about a multiplication so thick as to color the water. Not all colored waters are deadly, however, for nonpoisonous species can undergo population explosions too, as can other kinds of brightly colored organisms.

The culture of algae in laboratories takes little space and large numbers can be grown in a short time. They are much used to study processes that take place in green plants, and have been especially useful in analyzing the steps in photosynthesis. Currently experiments are being carried out with algae to devise closed ecological systems—self-sufficient "worlds"—to provide man with food and oxygen during space travel. So far no such system has been created that can support a man, but a small model was built that kept a mouse alive for six weeks. The system consisted of a closed container, lighted and with the temperature controlled. The mouse lived on a support above a tank of water in which nutrients were dissolved and which contained algae and bacteria. The algae made food and released oxygen. Periodically some of the algae were filtered out and automatically transferred to the mouse's food dish. The algae utilized the carbon dioxide released by the mouse, and the mouse's droppings were decayed by the bacteria to give additional nutrients for the algae.

People seeking ways to feed the world's starving millions see a solution in the wealth offered by the ocean. It is now possible to make a tasteless and odorless protein concentrate from algae, and, as a matter of fact, from yeasts and fungi as well. Perhaps it will be used first to feed domestic cattle, supplementing available grass. In time and with education it may be accepted by human beings as an additive to accustomed foods. Thus, for the first time man will be able to bypass the long and particularly wasteful food chains and partake directly of the basic foodstuff of the ocean.

chapter four

Viruses and Bacteria

VIRUSES are the smallest of the earth's organisms yet discovered and are among our worst enemies, causing many diseases of man, green plants, and animals. All of us have harbored viruses that have resulted in suffering; nearly everyone has had the common cold, influenza, measles, chicken pox, mumps, and a wart or two. Most of us fortunately escaped smallpox through vaccination, poliomyelitis through good luck or immunization, and rabies partly because it is more rare and partly because we protect ourselves by vaccinating our pets. At times, rabies is prevalent in wild animals.

Plants, both cultivated and wild, are susceptible to virus diseases such as curly top of sugar beets, peach yellows, dahlia stunt disease, leaf roll of potatoes, and tobacco mosaic disease. There seem to be several strains of the latter which can infect many other plants. In many plant virus diseases the foliage is mottled or streaked green and yellow. The flowers have irregular breaks in the color and may be deformed as well. Infected plants gradually degenerate. The well-known "parrot" tulips and some other variegated flowers are actually infected with a virus which has made a kind of "truce" with the host plant.

Virus diseases of plants cannot be cured. By the time the symptoms show, the virus has traveled all through the plant's system. Removing streaked leaves, for instance, will not free the plant from infection.

But by a very elaborate and specialized kind of tissue culture it is possible to free some kinds of plants from virus.

The first inkling that agents smaller than bacteria could cause disease was obtained in 1892 by a Russian botanist, Dmitri Iwanowski, from studies with the tobacco mosaic disease. He extracted sap from diseased plants and forced it through a filter so fine that not even the smallest bacteria could pass through it. The bacteria-free extract was then inoculated into healthy tobacco plants, which then developed the same symptoms. Clearly, agents smaller than bacteria were present in the sap, although they could not be seen with the strongest light microscope. They were given the name of "virus." Since then much has been learned about them.

In recent years scientists have been able to see these mysterious viruses for the first time by means of photographs made with the electron microscope. In contrast to the light microscope, which enlarges 3,000 times at best, the electron microscope yields magnifications of 250,000 times. Now that many viruses have been isolated and photographed, they prove to be quite varied in size and shape. The polio virus is small and spherical, rather like a cube with its edges and corners rounded. Tobacco mosaic virus is a long cylinder. One that causes the common cold is spherical and much larger than the polio virus. There is a type that attacks bacteria, called a bacteriophage (bacteria-eater), that has a prismatic head attached to a cylindrical tail.

Oddly, viruses have characteristics of both living and nonliving matter. They can be crystallized. This was demonstrated in 1935 with the tobacco mosaic virus, and the crystals, when inoculated into healthy tobacco plants, reproduced the mosaic disease. Viruses can be purified by chemical techniques. They cannot be grown on the usual culture media, but must be grown in living tissues. (Bacteria, by contrast, are independent—given a source of food and moisture, as on agar culture media, they can use the food and reproduce.) Viruses are now known to have a more complicated anatomy than was first thought, a structure that puts them in the category of living things. They have a core of nucleic acid, the carrier of the genetic code, which is surrounded by a protein coat. In most of those that cause human disease the nucleic acid is DNA, but in those that cause mumps and influenza and most plant virus diseases it is RNA. When a virus invades a living cell, its DNA or RNA takes control of the cell and directs it to manufacture more viruses. This ability to bring about its own reproduction is another evidence of the virus's belonging to the realm of the living. It is possible that early

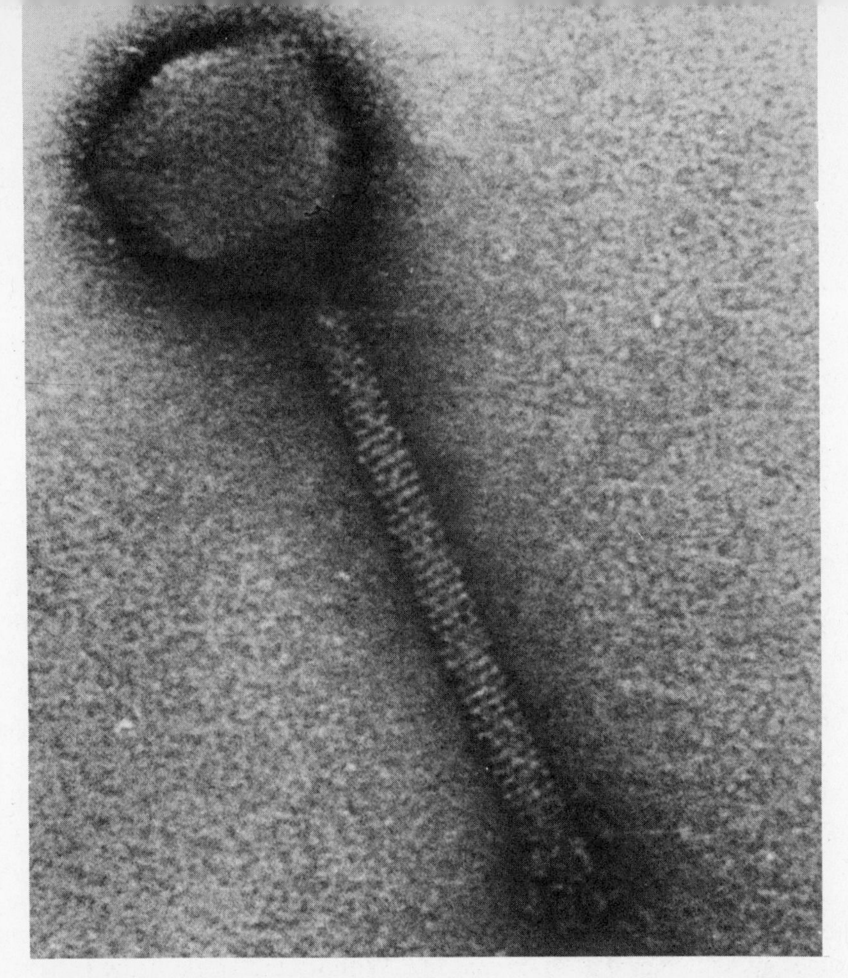

A virus that infects bacteria, known as a bacteriophage, magnified 400,000 times. (*A. K. Kleinschmidt*)

viruses were independent organisms which lost many of their characteristics upon assuming a life of parasitism. It is also possible that viruses evolved much as they are today, having changed little in their long history.

The momentous discovery that nucleic acid is the carrier of the genetic code, which led in turn to the great revelations of modern genetics, was made from a study of viruses that parasitize bacteria, the bacteriophages. The virus was seen to attach itself to a bacterium by means of its tail. It then shed its protein coat and sent its nucleic acid molecule into the bacterium. Many new units of the virus were quickly formed within the bacterium. They then erupted through the cell membrane of the now dead bacterium and repeated the process with other bacteria. This was a clear demonstration that the nucleic acid carried the necessary information for duplication of the virus, that it had the power to turn off the DNA of the bacterium so that its own DNA could direct the protoplasm of the bacterium to manufacture more molecules of virus nucleic acid and the protein coating material, and that it could govern their assembly into intact units of virus.

Like other organisms, viruses undergo mutations and thus become adapted to previously immune hosts and to altered living conditions. Their rate of mutation is exceedingly rapid. For example, the influenza virus has mutated a number of times, each time becoming capable of infecting people who had built up immunity to the older forms. In recent years alone, we have seen the Asian flu superseded by the Hong Kong type.

Things in nature did not evolve in neatly compartmentalized groups. It is man who wishes to put everything in categories. When something doesn't quite fit, he has to decide to what other kind it is most closely related. The viruses are quite clearly set off by themselves. But there is a small group called the rickettsias that are halfway between the viruses and bacteria. They look like small rod-shaped bacteria, but in common with the viruses they cannot be cultured apart from living cells. They live on lice and ticks and can be transmitted from them to man, in whom they cause typhus, undulant fever, and Rocky Mountain spotted fever.

For every green plant we can see, there are billions of nongreen ones, the fungi, 80,000 to 100,000 kinds in all. Their total mass is twenty times greater than that of animal life. Without the work of the beneficial ones continued life on earth would be impossible, but unfortunately there are members that also are extremely injurious.

Bacteria are the simplest of the fungi. Since they reproduce by splitting in two, they are called the fission fungi, the others being classed as true fungi. Bacteria live everywhere—in the soil and water, on and in animals, and on plants and sometimes within them. They are very ancient and primitive, and probably evolved about the same time as the algae.

Although they have not changed structurally for two billion years, they have changed physiologically, adapting themselves to new sources of food whenever such became available. The bacterium illustrated here lived two billion years ago, and in form and appearance resembles present-day bacilli. It was found fossilized in chert of the Gunflint Iron formation on the north shore of Lake Superior. Its internal structure and mode of reproduction were also probably similar to those of

Two billion years ago these bacteria were active. They were preserved in chert in the Gunflint Iron Formation near Schreiber, Ontario, and are similar to modern bacteria. (*E. S. Barghoorn*)

modern types. Some of the earliest bacteria made their own food, and some obtained it from living or dead algae, which were the only other organisms on earth at that time. As millennia passed and living things became more and more complex, some bacteria preyed on the new forms, while others used their dead remains for food. Still others established mutually helpful relationships with both plants and animals.

All bacteria are one-celled and are quite simply constructed, with little internal specialization. They have no nucleus. A cell wall surrounds the living contents, in which there occurs one or more strands of genetic material, the DNA. The DNA transmits information from parent to offspring, as it does in green plants, animals, and man, with of course a different arrangement of molecules to carry its code. Such bacteria are easily cultured in the laboratory on an agar gel containing nutrients, where they form characteristically shaped colonies of various colors, some very beautiful. Individual species can be isolated by reculturing techniques.

As the bacteria are separated from the true fungi, so are they themselves classified into various groups, making it difficult for the nonspecialist to keep them straight. However, three of these groups play important roles in our lives—the true bacteria, the actinomycetes, and the spirochetes—and they and the problems they cause appear frequently in the news and in conversation.

Most bacteria are nongreen plants, containing no chlorophyll, and secure their food from dead organic matter or from a living host. A few have bacteriochlorophyll and carry on photosynthesis. Still another few carry on chemosynthesis, using energy obtained by the oxidation of such substances as sulfur, ammonia, nitrite, and hydrogen. Among the smallest of plants, they measure only one ten-thousandth of an inch across. Five hundred billion would weigh only one-thirtieth of an ounce. They come in three shapes—spheres, rods, and spirals, technically referred to as the coccus, bacillus, and spirillum forms. Some are motile; the ability to move is rare in coccus, common in bacillus, and universal in spirillum. The spirillum forms are long and slender with a corkscrew shape, and they wriggle along like small snakes. Some have flagella to propel them. The rod-shaped bacilli also have flagella; some have only one, like a tail, but others display them at both ends or all over their surface.

When water is lacking or conditions become unfavorable, many species (mostly the soil-dwelling ones) form spores. This is not a means of reproduction, as it is in other plants, but merely a technique for survival. The protoplasm loses water and becomes dense and a heavy wall is formed around it. The original membrane of the bacterium ruptures, and the spore is released. Spores are extremely resistant to heat and cold; some can withstand boiling for several hours, and a few can survive cold down to −454° F. When spores come to rest in favorable places, they germinate and develop into active bacteria.

The reproductive potential of a bacterium surpasses that of most other organisms. It can split in two every twenty minutes. As one divides, its cell content, including its DNA, is equally divided between the two resulting cells, and a new cell wall is formed between them. A single organism can become two in twenty minutes, four in forty minutes, eight in an hour. If food were plentiful and no obstacle presented itself to continued division, these bacteria could double their number every twenty minutes indefinitely. There would be nearly a billion in 10 hours, and a trillion billion in 24. Those who like to think in terms of objects placed end-to-end or covering acres might enjoy the figures arrived at by a bacteriologist: he calculated that starting with one bacterium in milk at room temperature (and assuming that the rate of division remained constant and the supply of milk did not give out) at the end of thirty hours there would be enough bacteria to fill 200 five-ton trucks! The blessing of refrigeration can be noted here, for at temperatures just above freezing, it takes six hours for a bacterium to achieve one division. Actually, unlimited division rarely takes place for any length of time. In milk, for instance, a product of the bacteria is lactic acid, which sours the milk and eventually slows down the rate of division. Like a human city, a bacterial culture can poison itself with its own waste products.

One group of species prefers cold; they live in the deep ocean, in cold springs, and in cold soils. Another group thrives best in temperate areas. Still another group is adapted to withstand heat up to 167° F., and they live in hot springs, manure piles, compost heaps, and haystacks. Spontaneous combustion of hay often occurs when heat from the bacteria raises the temperature to 160° F. Subsequent oxidation of the products they manufacture raises the temperature further to 400° F., the point of combustion. Ironically, the hay must be slightly damp before this incendiary process can begin!

Although there is no sexual reproduction in bacteria as we know it in higher organisms, they do occasionally exchange genetic material by conjugation. Two bacteria come together and form a protoplasmic bridge. Through this channel cellular material flows from one to the other. They then separate and each continues to divide by fission. Experiments have been carried out on growing two strains of bacteria together, with the result that offspring develop with characteristics of both types, showing that they do indeed exchange genetic material.

Bacteria that feed upon living organisms are parasites, some of which are responsible for various diseases. Some of the coccus forms stick together in various ways. Some travel in pairs in a form called diplococcus, and one species of this form causes lobar pneumonia. Some link up to form chains, a form known as streptococcus; they cause sore throat ("strep throat"), rheumatic fever, and scarlet fever. A form that resembles a bunch of grapes is known as staphylococcus, and species of this type cause boils and other pus-forming infections, and osteomyelitis, a bone disease. The rod-shaped or bacillus forms cause typhoid fever, tetanus, and tuberculosis. Among those with the spiral form is one that causes cholera.

A type distinct from these is the spirochete, a long, motile one, that includes species that cause syphilis, trench mouth, and yaws.

Three types of food poisoning are due to bacteria. Botulism is caused by a toxin given off by a bacillus that can live without oxygen, *Clostridium botulinum*. Even the smallest taste of the toxin can be fatal. These

Roots of a pea plant. Bacteria which grow in the nodules (swellings on the roots) fix nitrogen. (*O. N. Allen, University of Wisconsin*)

anaerobic bacilli have spores that are not killed by boiling, and when the spores are present in meats or nonacid fruits and vegetables canned by the boiling method, they survive to germinate and multiply in the jars. The toxin they secrete accumulates. When the foods are merely warmed before serving, the toxin is not destroyed. But since it is a protein, it can be broken down by fifteen minutes of vigorous boiling, and the foods thus made safe for eating. The spores do not germinate in a strongly acid medium, so acid fruits like peaches are generally safe.

Another kind of food poisoning is caused by a staphylococcus, formerly called *Staphylococcus aureus* and now called *Micrococcus pyogenes* variety *aureus,* introduced into food by unsanitary hands or utensils. This is the aftermath of many a picnic or community meal where food is left standing for long periods without refrigeration. The bacteria multiply with speed in warm cream-filled bakery products, potato salad, poultry stuffing, and gravies. Again, the agent is a toxin produced by the bacteria, and while it is not fatal it can cause acute and severe digestive upsets. (This was once called ptomaine poisoning but the *ptomaines,* although products of putrefaction, have been found *not* to cause poisoning when eaten. The name was wrongly used and unfortunately stuck. No new name has been coined.)

A third kind is called salmonella poisoning, after the name of the bacillus that causes it. In this case it is the *Salmonella* bacilli themselves that are injurious; they multiply in the intestinal tract after food contaminated by human or insect carriers is eaten.

Plant diseases are caused by many species of bacteria. They enter the plants through the leaf pores—the stomata—aided by the presence of water on the foliage, or through wounds. Bacteria can be transferred from a diseased plant to a healthy one by contaminated tools or by insects. Diseases both minor and catastrophic are brought about by different species—ring rot of potatoes, bean blight, alfalfa wilt, fire blight of apples and pears, soft rots of various vegetables and of orchid and iris rhizomes, and many leaf spots. Many of the diseases are specific, that is, certain bacteria will attack only certain plants. When one of these crops is grown in the same place for several years the bacteria build up and the disease becomes more severe. Crop rotation is therefore helpful in controlling diseases where bacteria remain in the soil. Planting a crop that is resistant to the bacteria in that spot allows the population to decline to the point where it may be possible to grow the original plant again. There is a host of fungicides now available with which to battle plant diseases; and a continuous search goes on for plants resistant to the diseases that commonly attack their kind.

There are other bacteria that live in partnership with man and animals, and since their activities are mutually beneficial they are said to be symbiotic, which means living together. For example, those that live in the rumen of grazing animals first turn the cellulose of the grasses into sugar and, by adding nitrogen, form amino acids. As the bacteria pass on through the alimentary canal, they die, and the amino acids, proteins, and vitamins B and E which they contain (not available from the original fodder) are used by the ruminant.

Bacteria that live in soil and water and consume dead plant and animal material are saprophytic. They digest the compounds manufactured by organisms during their lifetime and thus reduce their remains to chemicals that can be used again.

Among the soil dwellers are the actinomycetes, which have thread-like bodies similar to the mycelium of fungi, yet lack nuclei, as do the true bacteria. They include several species of *Streptomyces* that have given us the antibiotics streptomycin, Chloromycetin, Aureomycin, and Terramycin. Scientists are studying soil samples from all over the world, hoping to discover others that will give additional antibiotics and possibly antitumoral agents. A few actinomycetes are injurious, such as the one that causes the scab disease in potatoes.

Green plants cannot use pure nitrogen; it must be in the form of certain compounds—ammonium nitrate, for example. Several species of bacteria have the ability to fix nitrogen, as the process is called. Two kinds, *Clostridium* and *Azotobacter,* that live in the soil contribute about 25 pounds per acre per year, making fallow lands more fertile with respect to nitrogen. More efficient than these are several species of *Rhizobium* that live on the roots of legumes, members of the pea family, in a symbiotic relationship. The bacteria obtain some nourishment from the legumes, and in turn form nitrogen compounds on which the plants thrive. Often farmers plant a legume crop in alternate years with some other crop for the purpose of fertilizing the soil with nitrogen compounds.

Earth's sanitary engineers are the saprophytic bacteria and fungi. They are also the conservationists. The supply of minerals with which the earth has been endowed is limited. Little comes to us from elsewhere in the universe, through the slow fall of meteoric material. But the water on our globe has been used and reused for millions of years; so has practically everything else. The minerals in our bodies were once part of dinosaurs and ancient plants. The air we breathe was inhaled and exhaled by primitive man. Yet no minerals or oxygen would be left for us if they were still locked up in the bodies of earlier creatures, if bacteria and fungi had not removed them and returned them to the earth, year by year. Nor would the face of the earth be a very pleasant place to live if the sanitary workers should quit their job. Then not a leaf or banana peel or tree trunk, not a fly or horse or man would turn to dust. Instead the earth would be covered with piles of corpses, dead plants, and other refuse, mummified perhaps, but intact. The picture of a big city in the throes of a garbage collectors' strike gives a mere hint of the condition that would prevail. When the last vestiges of minerals became locked up, all living things would starve, plants as well as man.

The bacteria and fungi are thus nature's balancers, keeping the wealth of the earth in circulation. As soon as an organism expires, they return to the soil the minerals of which it was composed, and they return carbon dioxide to the air. Plants reuse the minerals and the carbon dioxide, and animals feed on the plants. The cycle comes full circle when bacteria and fungi again consume the bodies of the plants and animals.

The population of bacteria and fungi usually remains in equilibrium with the amount of organic refuse. When refuse—their food—is scarce, they do not multiply rapidly; when food is abundant, the population explodes. However, they cannot do much with old bottles, tin cans, bedsprings, broken-down cars, and plastic toys. We have to take over this cleaning-up responsibility ourselves if we are to maintain a beautiful landscape.

The contrast is striking between the clear water in remote mountain streams and the malodorous brownish-purple fluid flowing in rivers passing by our cities. Once, all the water flowing in this country was clean and palatable. When the Indians held the land, and even when the first settlers came, the small amount of refuse that entered the streams was not too much of a task for the beneficial bacteria to handle. It takes only time to reduce the refuse to pure compounds. Today such huge quantities of sewage, offal from slaughterhouses, and poisonous chemicals are dumped into the rivers that the bacteria cannot cleanse the water in the short time it takes a stream to flow from one city to the next. Each city has to use the polluted water from its upstream neighbor. Large water purification plants are necessary to treat the water before it is offered to the inhabitants. Many sewage-treatment plants are inefficient, and in addition a great deal of material is still dumped directly into the water by factories and various types of industrial plants, sometimes against the law and in the dark of night.

Given time, bacteria will transform organic refuse into harmless substances, and if the job isn't completed in the rivers, it will be finished by bacteria in the ocean. But time is lacking in our modern cities. The job must be done more quickly. Man must find ways of aiding the bacteria.

A bigger problem now has come into being, that of the disposal of chemicals that bacteria cannot reduce. Detergents were hailed as a wonderful new development until it was discovered that bacteria could not break down the formulas first put on the market. Biodegradable detergents—those capable of being reduced—had to be invented. But it will be some time before those earlier detergents with their still-present suds are washed out of the streams, the wells into which they have seeped, and the land on which the contaminated water was applied.

More dangerous chemicals than these are creating the supreme problem of our day—DDT and chemically similar insecticides and radioactive wastes, which neither man, nor animals, nor plants can change. Only time can lessen their lethal activity. In the meantime, they travel from one organism to another, and even the most remote inhabitants of the earth are not immune.

chapter five

True Fungi, Slime Molds, and Lichens

SINCE the dawn of history fungi have been wrapped in superstition, associated with sprites, sorcerers, and little people, and linked to magic and religion. It was quite natural that these plants that exhibit so many strange qualities—they reveal no roots or leaves and pop up out of the ground overnight—should be treated with awe and wonder. Imagine entering a secluded forest glen on a moonless night and finding mushrooms glowing with their own eerie light! You could easily picture elfin creatures dancing in the glim, their king and queen, perched on the largest and brightest mushrooms, ordering tricks to be played on human beings. The sudden appearance of a perfect ring of mushrooms in an open field would not be difficult to interpret as the scene of a fairy carnival the night before. If you felt close to your gods and your ancestors, and were to eat certain mushrooms your tribe held sacred, you could easily interpret the strange dreams that ensued as visions and signs, and attribute magic powers to such miraculous plants. Although today we are inclined to discount magic properties, the fungi remain marvelous enough, even more so than early man could know.

Beautiful to our eyes are the mushrooms of field and forest, and the shelf fungi that grow on trees and rotting logs. Equally beautiful under the microscope—especially when cultured on nutrient agar in shallow glass dishes—are some of the hundreds of soil fungi, which yield colonies ranging in color from white through yellow, blue, red, violet, and green. Many fungi, among them those we can eat and those from which we obtain antibiotics, are saprophytic and serve (along with many other kinds) as scavengers of dead organic material, bringing about decay and returning minerals to the soil and carbon dioxide to the air. Some live as parasites on plants, animals, and man, and these cause personal suffering as well as losses in both crops and livestock. Still others do us no harm if we let them alone, and are just as useful ecologically as their relatives, but are poisonous to eat.

The true fungi are far more complex than the bacteria. Their cells have specialized structures such as those found in higher plants—a nucleus, mitochondria, and ribosomes. Most of them have a cobweblike body known as mycelium, whose threads penetrate dead organic matter or—in the case of parasitic kinds—the tissues of the host. If you pull apart a rotting log, you can see the saprophytic mycelia of several species growing everywhere within it. You can observe the parasitic mycelium of powdery mildew spreading a white pallor across the leaf surfaces of infected plants. A few fungi do not produce a mycelium, among them the one-celled yeasts, the most familiar of which are those used in bread-making and in the brewing industry.

A fungus infected and killed this darkling beetle. The stalk with the capsule at the top is the reproductive structure of the fungus. (*Gerald Lang*)

Fungi have a more complicated system of reproduction than bacteria. They reproduce by spores, some of which are the end products of sexual reproduction, others formed without this process.

On the evolutionary ladder the most primitive fungi were aquatic and produced swimming spores. The water molds (species of *Saprolegnia* and their relatives) still do. Some of these are saprophytic and live on dead plant and animal material. You have probably seen their little pincushions or ruffs of delicate white threads bristling out of seeds, dead insects, or bits of refuse floating in ponds or slow streams. A few species are parasitic on fish and become troublesome in hatcheries and aquariums. Perhaps you have seen one growing as a feathery tuft on the nose of a goldfish.

The spores of the fungus that causes chestnut-tree blight, *Endothia parasitica,* rely on insects, squirrels, and birds for their dissemination. The disease, introduced from China, has killed nearly all the chestnut trees in the United States. The spores are embedded in a sticky substance that smears and adheres to the feet of animals. In one instance, a billion spores were washed from the feet of a single woodpecker.

The shotgun fungus depends on horse dung for its growing medium and uses a peculiar technique for insuring that its spores will find sustenance. The fungus, *Pilobolus,* grows on the dung and forms spore cases at the top of stalks. Behind each spore case is a little bulb filled with fluid. The spore cases mature during the night. As the sun comes up, each orients itself toward the light, and the bulb gives a sharp contraction, shooting the case containing the spores in that direction. The effect is to send the spore case away from the parent colony so that it may land on fresh grass and be eaten by a horse. In the alimentary tract of the horse the spores are released from the case, and they emerge with the dung, on which they germinate and grow.

Most spores, however, are not so discriminating and are borne by the wind. The air is full of them at all times, and the wind carries them hundreds of miles, often at great heights. In fact, the fungi have solved many of the problems that plague man in his exploration of space. The spores are so nearly weightless that they can practically defy gravity. Because of their negligible mass they are completely free-moving, able to whirl in one direction at one moment and shift to another direction the next. Their "riders" are in a state of suspended animation, oblivious to the lack of water and food, able to withstand extremes of temperature, unaffected by rarity of oxygen. Some can travel for months.

When they descend to earth, if they happen to land in a spot that gives them congenial conditions, they resume an active state, germinate, and grow.

In outward aspect some could almost be models for spaceships. Particularly beautiful and well adapted for such high atmospheric ballooning are the spores of the green mold *Aspergillus*. Two that resemble miniature Saturns (*Aspergillus nidulans* and *quadrilineatus*) have two or four rings layered around their equators, which add to their soaring power. Another (*Aspergillus variecolor*) has two flat six-pointed stars girdling its middle. One that looks like a walnut out of its shell (*Aspergillus rugulosus*) also has rings to give it a lift. These are but a few of the shapes among the thousands of kinds traveling the airways. The spores of mosses and ferns accompany them and, in season, pollen grains are also present in tremendous numbers.

By using the wind for transportation the fungi are able to spread far and wide, and they are not barred from any spot or any substance upon which they can live. Their activities are especially evident in damp climates, where their spores land on and attack shoes, food, wallpaper—in fact most human possessions of organic material. Wherever living things dwell, the fungi will waft in sooner or later to make use of their refuse and their remains.

Space travelers of the plant world have been riding the air currents for millions of years. (*Drawn by Abbie M. Current*)

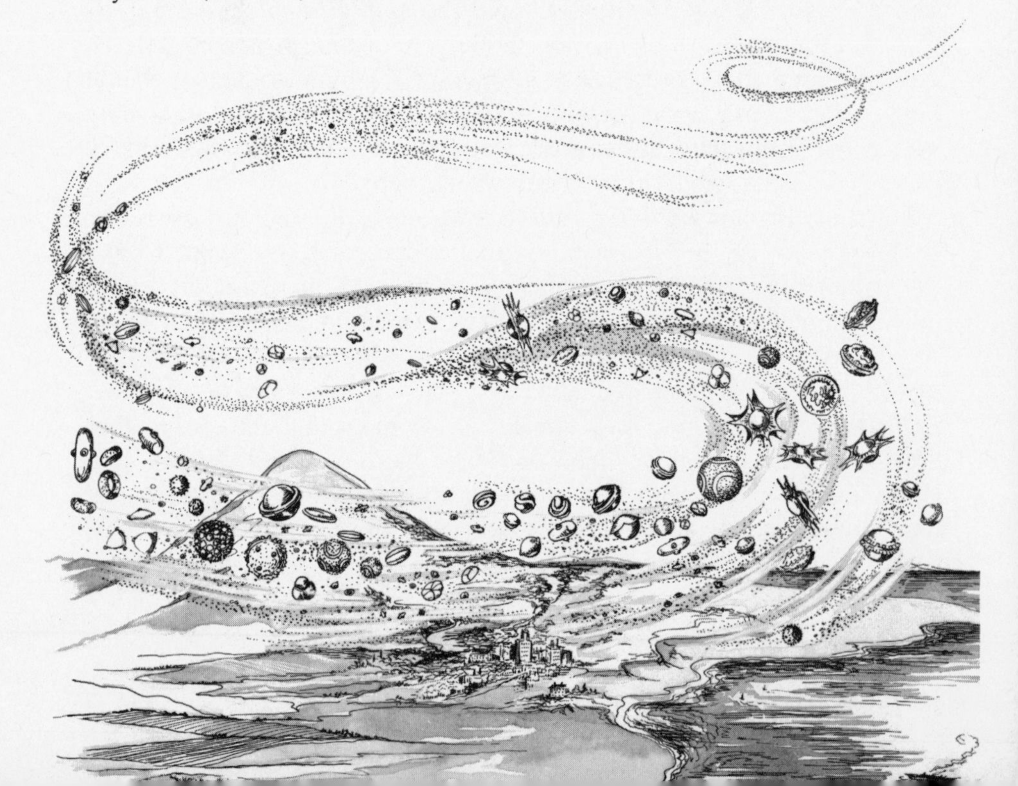

For all their striking appearance, the familiar mushrooms are merely the reproductive structures of certain soil-dwelling fungi. The mycelium grows hidden in the soil, foraging on decaying leaves, rotting wood, manure, and other organic matter. After the mycelium has grown sufficiently, and after it has gotten the necessary amount of rain, it forms small buttons which increase in size and appear above the soil surface, where they develop into the characteristic mushrooms. Each has a stalk that supports a cap with gills on its undersurface. These gills produce enormous numbers of spores, as many as two billion to one mushroom. In order for the spores to fall free, the gills must be in a vertical position; a mere ten-degree tilt causes them to adhere to the gills. A mushroom will usually orient itself so that the cap has a tilt of less than two degrees. When ripe, the spores are popped off the gill with just enough force to move them into the middle of the narrow space between gills. Then, unencumbered, they drop. In still air they settle to the ground; when a breeze is stirring they are blown away.

Beautiful spore prints can be made by setting the caps of mushrooms on paper and leaving them overnight. The spores fall on the paper, leaving an almost photographic picture of the design of the cap and the pattern of the gills. Some spores are white, and these show up best on dark paper; some are of lovely colors that blend or contrast with papers of various hues. (Be sure to spray the finished print with artists' fixative, or it will smudge and be ruined.) The color of the spores is one of the features by which species can be identified.

The mycelium can grow in the ground for many years without producing reproductive bodies. The temperature and amount of rain must be just right, and the mycelium must be vigorous, before mushrooms are formed. In some areas these requirements are met regularly, in others only occasionally. One summer in a particular area of the Rocky Mountains, several kinds of mushrooms appeared that had not been seen for decades, and those kinds that normally grew in small quantity sprouted in fantastic abundance. Residents of the area refer to it as "the year of the mushrooms." Evidently soil temperatures milder than usual, combined with melted snow and subsequent rains, had produced just the right conditions.

Mushrooms fascinate us with their variety of size and delicacy of structure. From tiny little forms so fragile that the caps break from the stalks at the slightest touch, they range up to huge 6- and 8-inch kinds that small animals can sit upon (hence the nickname "toadstool"). Some grow even larger. Their colors go from ghostly white to cream, yellow, pink, orange, red, purple, brown, and near-black. The caps of some are nearly flat with the edges barely turned under, while those of others are shaped like cups or thimbles, Tyrolean hats, or even umbrellas turned inside out. The surface of the cap varies in texture from absolutely smooth—even slippery—to velvety, scaly, or warty.

"Fairy rings" are formed by mushrooms whose mycelium grows outward in a radial manner, like the ripples in a pool. Perfect rings are more likely to occur in lawns or pastures than in the forest, where organic matter is not deposited so evenly. As the mycelium spreads at its outer edge, it may die in the center, sometimes leaving the grass in the center also dead. Often the grass above the living mycelium is a darker green than elsewhere, making the ring quite noticeable. When conditions are right, mushrooms appear at the outer edge, arranged in a circle—the fairy ring. Four-leaf-clover designs can result when spores start off on their own and produce new, interconnecting mycelia.

"Fairy light," a faint luminescence that can be seen only on dark nights, is given off by several kinds of mushrooms; in some by the mycelium, in others by the cap or the gills, and only when these parts are young and fresh. It is the honey mushroom (*Armillaria mellea*) whose mycelium in rotten wood gives the yellowish or greenish light also known as "fox fire." It is said that woodsmen once used pieces of the glowing wood to mark their trails. The gills of the jack-o'-lantern (*Clitocybe illudens*) and the cap of the American form of *Panus* (*Panus stypticus,* so named because of its astringent nature) give off a greenish white light. Oddly, the European *Panus* is not luminescent.

Unfortunately, not all mushrooms are edible—some members of largely edible genera are deadly poisonous, and there is no simple rule to distinguish between them. It is not true that the brightly colored ones are poisonous and the dull ones edible, nor can they be told apart by the ease with which the cap can be peeled or whether they will turn a silver spoon black. The only safe way is to know a species before eating it. Of course, safest of all is to buy them in the grocery store, for the commercial ones are raised by hand. It would be a shame to avoid all wild ones, however, for some are truly delicious, their flavor and texture subtly different from *Agaricus campestris,* the meadow mushroom grown commercially. There are a few that can easily be recognized with practice, and experts in the region where you wish to gather them should be consulted. Some that an amateur can come to recognize are the oyster mushroom (*Pleurotus ostreatus*), the shaggy mane (*Coprinus comatus*), the delicious lactarius (*Lactarius deliciosus*), and, of course, the meadow mushroom just mentioned. Books are a great help, although the astonishing number and variety of species they reveal can be bewildering. Your first attempts to identify wild mushrooms by means of keys and pictures will enlighten you as to the difficulties involved. So many species are there that an expert often deals only with those found in one region, or only with certain groups; when presented with a kind out of his field, he will not care to identify it with certainty.

As to the edibility of the species, some are clearly stated to be "deadly poisonous," "poisonous," or "edible," and of those that are designated edible, the flavor too is sometimes described. However, even the most daring and willing novice will be deterred by such phrases as "edibility unknown," "edible, but use caution" (what does *that* mean?), "edible, but in some people may cause nausea or diarrhea," "not poisonous, but has a disagreeable flavor," "edible when young" (*how* young?), "at times poisonous," and "edibility in debate" (this poses all sorts of questions!). Incidentally, of the luminescent mushrooms, the jack-o'-lantern and the *Panus* are poisonous, whereas the honey mushroom is edible. No wonder fungus poisoning was a major cause of death to gourmets in Imperial Rome!

Ninety percent of the deaths brought about by mushroom-eating are caused by species of *Amanita*. All but a few of them are poisonous to some degree, and two are deadly—the fly amanita (*Amanita muscaria*) and the destroying angel (*Amanita verna*), the latter the most deadly of all mushrooms. The top of the cap of the fly amanita is usually bright orange or red, flecked with raised white specks, whereas the destroying angel is pure white with a smooth and sticky cap. Both

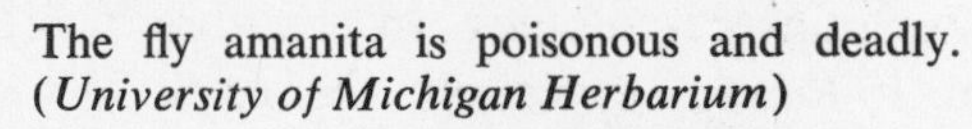

The fly amanita is poisonous and deadly.
(*University of Michigan Herbarium*)

species have a ring on the stalk below the cap, a cup at the base of the stalk, and white spores. In both, the gills are white, although they turn yellow with age in the fly amanita. (Note: some of these features are found in other mushrooms, but the amanitas are the only ones that have *all three*—the ring on the stalk, the cup at the base, and white spores.) The toxic substance of the amanitas attacks the central nervous system. Symptoms of poisoning include profuse perspiration and salivation, diarrhea, delirium, convulsions, and paralysis of respiration. The destroying angel is so poisonous that you should not even touch it for fear that a particle of it might reach your mouth. The fly amanita is equally poisonous, but in quantities larger than a taste—although just how large a dose is fatal is not known and may vary with the individual. Long ago, Europeans used it to kill flies, hence its name.

In minute amounts, the fly amanita is hallucinogenic, and was used for this purpose by tribes in Siberia and by the Lapps in northern Europe. Needless to say, experimentation is not recommended! It is reported that the Berserkers in Norway may have induced their fits of savage madness by eating the fly amanita. Our phrase "gone berserk" comes from this source. A person who eats a bit of dried fly amanita or drinks an extract undergoes a period of twitching and trembling, followed by visions of the supernatural and illusions of grandeur, and often a desire to confess his sins. Occasionally he may become violent and dash around madly until he falls exhausted into a deep sleep. The hallucinogen contained in the mushroom is excreted from the body unchanged, and the people of the tribes who used it found that they could reintoxicate themselves by drinking their own or another's urine.

In other parts of the world, far different species of mushrooms are partaken during religious ceremonies. In the Huatla region of Mexico certain species of *Psilocybe,* especially *Psilocybe mexicana,* are used by the Indians in ceremonies that play an important role in their lives. The tradition has been carried on for generations. After the sacred mushrooms have been cleaned and passed through burning incense, they are apportioned to the participants, accompanied by prayers. The room is then darkened. Soon nausea comes, and the beginning of visions which continue after the nausea passes. Chanting accompanies the rites, which are carried out with decorum. Evidence of an ancient mushroom cult in Guatemala is seen in the many stone carvings of mushrooms, dating from 3,500 years ago. A human figure is represented in some of the carvings, sometimes shown in an attitude of dreaming.

Mushrooms held sacred by the ancient Indians of Central America and used in religious rites were artistically sculptured. This "mushroom stone" was found near Guatemala City and dates from 850 B.C. to 300 B.C.
(*Owned by Dr. Carlos García of Guatemala. Photo loaned by the Metropolitan Museum of Art*)

A coral fungus, *Clavaria pyxidata,* an edible species. (*Department of Plant Pathology, Cornell University*)

Rather formidable looking is the hedgehog or toothed mushroom (*Hydnum*). Most of the species are said to be edible, but one would surely have to try them in their young stages, for they become leathery or woody as they mature. They have spines, rather than gills, on which the spores are formed. The coral fungi (*Clavaria*) look nothing like their mushroom relatives, but more like marine sponges, with their fingerlike branches. The edible ones are among the most delicious of all.

A group called pore fungi have among their members *Boletus,* which has a typical mushroom shape with cap and stalk. But instead of coming from gills, the spores are formed in a tissue lining the underside of the cap that consists of channels with pores opening to the outside. In some the pore-bearing tissue is white, in others yellow, gray, or brown. Many are very good to eat, but those with red pores are poisonous. Once the edible kinds have been pointed out to you, they are not hard to recognize. The shelf fungi that grow on trees and logs are also pore fungi, and although there may be a really edible kind, most are unpalatable because of their tough, woody texture. They are beautifully decorative and of warm colors, and since they live on the wood, they may be used in arrangements with dried flowers. Some shelf fungi have a delicate undersurface that turns dark at the slightest pressure. They are living slates that can be used for writing or drawing, and later dried to preserve the inscription.

All puffballs are edible in their young stages. They consist of a sphere with a rind on the outside and spore-bearing tissue within. When they are young, both the rind and the inner tissue are white, firm, and tender. But when the inner tissue becomes brown and the spores begin to form, they are not good to eat. Finally, the rind becomes tough and dry; it then shrivels until at last it breaks open and releases the ripe spores. To a child on a walk in the woods, it's a great joy to find puffballs ready to release their spores. Stamping on them or squeezing them makes their spores spurt out like puffs of thick smoke.

The number of spores formed by a puffball is astronomical. A giant one about a foot in diameter produces about seven *trillion*—very few of which, perhaps only one, will find conditions suitable for growth, while the rest go to waste. Externally, however, the buttons (young stages) of many mushrooms, including poisonous ones, resemble puffballs. Before cooking any white spheres you pick up, slice them open to see if they really are puffballs. If not, the lengthwise section will reveal a developing cap and stalk.

A beautiful relative of the puffball is the earthstar, also edible when it is young—that is, well before the outer rind splits open. The rind splits into several pointed segments, revealing the inner sphere of spore-bearing tissue, and producing a star with four or more points.

Morels, items of gourmet fame, are quite different in appearance from any of the forms described above. They have a stalk, on top of which is a cluster of pits with ridges in between. The pit linings are the tissues that produce the spores. They must not be confused with false morels, many of which are poisonous and should be avoided. The top of the latter is merely convoluted (wavy) rather than pitted.

Many true fungi are parasites, causing serious diseases of plants—blights, rusts, smuts, and others. The parasite lives on food provided by the cells of the host. (This is true of the parasitic bacteria, the viruses, and of some seed plants as well.) It may be that forms that are now parasitic evolved from some that were once free-living but which became specialized for a parasitic life. Also, whereas an ancestral fungus may have parasitized a variety of species, its various descendants may each have adapted to live on only one.

The host-parasite relationship is a close and rather sensitive one. The parasite must not be such a scourge that it drives its host species to extinction, for it would then deprive itself of its means of livelihood. However, no individual host lives forever, so the parasite must have a means of moving on to fresh hosts in order to perpetuate itself. Over long periods of time, a host species usually develops some defense against the inroads of a parasite, and it comes to tolerate the invading organism without becoming entirely debilitated. The parasitic species in turn develops some defense against the host's efforts to expel it so that it, too, can prosper moderately well. A dynamic balance is usually reached. An occasional mutation leads to an individual with complete resistance to a certain disease, and it can be cultivated and multiplied. Sometimes, however, the parasite eventually mutates, too, resulting in a new strain that can attack that species. Man must then search anew for other resistant individuals.

There are always some diseased plants in an undisturbed natural community, but their numbers tend to remain low and the injury relatively minor. This is partly because species that have survived through the ages have become fairly immune to the native disease organisms and are not greatly damaged by them, and partly because the diversification of species within a natural region does not allow specialized, one-host disease organisms to spread readily. Diseases become more prevalent when man disturbs the harmonious balance of the natural community, particularly when a single kind of plant takes over on logged-over or burned-over areas, or when a single crop is planted in a field or orchard. Foreign organisms—such as the chestnut blight—introduced to plants not resistant to them wreak havoc in both natural and disturbed areas.

In 1845 much of Ireland was one huge potato patch. Early in the summer of that year the fields were thrifty and promised a high yield, but by August every vine was black and dead. Panic-stricken peasants who dug for tubers found only rotten and inedible masses. The blight swept the country, and starvation followed in its wake. Because the Irish people depended so greatly on potatoes for food, the loss of the crop caused the starvation of a quarter of a million people. A million and a half survivors emigrated to the United States. People everywhere were aroused and demanded more information about the disease. They were answered with speculations—even that the weather was to blame—until the Reverend M.J. Berkeley carefully studied the diseased plants. He discovered that a fungus was invariably present on them, growing in leaf, stem, and tuber, and reproducing by spores that emerged through the leaf pores. Describing the fungus in detail, he advanced the revolutionary idea that it was the cause of the disease. This was the beginning of the science of plant pathology. Since then, our knowledge of fungus diseases has increased tremendously, and with this knowledge we have been able to control many of them.

A particularly devastating disease, and one which was especially puzzling until scientific detective work unraveled its secrets, was the white pine blister rust, *Cronartium ribicola.* Brought in from Europe, it attacked white pines in the east and then advanced westward, killing great numbers of trees. The discovery that led to its control was that the rust spent alternate generations on two different hosts, the white pines and either currants or gooseberries, species of *Ribes.* Spores produced on the pines cannot infect other pines, but they can and do infect *Ribes.* Through the summer, spores produced by *Ribes* infect only other plants of *Ribes.* Then in the fall the fungus on *Ribes* produces a different type of spore that moves back to the pines and infects them. Eliminating the *Ribes* plants, either by grubbing them out or by using weed killers, breaks the cycle and prevents the fungus from completing its reproduction. Several other rusts are known to require two hosts, among them the blackstem rust of wheat that spends the other part of its cycle on barberry, and a rust that moves between apple and juniper.

Another imported killer is the Dutch elm disease caused by *Ceratocystis ulmi.* Beautiful elms around homes, in parks, and in forests have been killed. The disease was first observed in 1930 in Cleveland and Cincinnati. Today it is found throughout much of the United States. Wilting and yellowing of the leaves are the first symptoms, followed by death of branches or the entire tree. The fungus depends on two insects—the native elm-bark beetle and the European elm-bark beetle—to transport the spores from one tree to another. If the beetles could be eliminated, the disease could be controlled. Many cities have sprayed the trees with DDT or other insecticides, but always enough beetles escape to keep spreading the fungus. Sometimes the cure has been worse than the disease, for the insecticides killed beneficial insects as well as birds, thereby upsetting the natural balance. When diseased trees are close to healthy ones, intermingling roots often form natural grafts, carrying the disease from one tree to another.

Oak wilt, another serious disease, attacks all species of oaks and has spread throughout large areas of eastern United States. The spores are carried by birds and insects, but they can infect a tree only through breaks in the bark. Careful pruning of infected limbs and removal of parts damaged in other ways can prolong the life of a tree.

Some soil fungi are parasitic on seed plants, and most gardeners and greenhouse growers are familiar with their depradations. Species of *Rhizoctonia* and *Fusarium* cause root rots. Species of *Pythium* also cause root rots, but are especially damaging in seed beds where they cause "damping off." In the latter the little plants suddenly fall over with the base of the stem rotted.

A mutually helpful relationship, called mycorrhiza, exists between fungi and the roots of many seed plants. The mycelium of the fungus forms a sheath around the tiny feeding roots and is believed to enable the plant to take up nutrients more effectively than they would otherwise be able to do. Some seed plants are so dependent on the fungus that they grow poorly when it is absent. The degree of dependence may vary with the environmental conditions, however. Kinds that have mycorrhizae are heathers, orchids, some pines, hickory, larch, birch, and cycad. Interestingly, among the fungi involved in the mycorrhizae of trees are such conspicuous genera as *Boletus, Amanita, Lactarius, Cortinarius,* and *Russula.* There are many mycorrhizal fungi that have microscopic reproductive bodies, and these can be detected only by study with a microscope.

Among the strangest phenomena of fungi are the slime molds, two classes of which exhibit quite different habits. One is a mass of gelatinous protoplasm containing nuclei but no apparent cell walls; the other consists of separate cells. For a plant, each leads a most remarkable life.

In its vegetative stage, the gelatinous slime mold can be seen on moist fallen logs and tree stumps, ranging from the size of a pinhead to 4 inches in diameter. The mass glides slowly along, engulfing bacteria, other fungi, and organic matter as it goes. It may be white, yellow, violet, red, blue, black, or grey. When a period of dry weather causes exhaustion of its water supply, the slime mold reproduces by forming delicate cases in which spores develop. Depending on the species, the spores may be stalked or unstalked, round or elongate in form. At maturity, the cases open and liberate the spores, which are scattered by the wind. When conditions are favorable—that is, when wet weather comes again—the spores crack open and each one liberates a swimming cell, known as a swarm spore. The swimming cells divide once or twice and then fuse in pairs to form new cells, each of which reverts to the slime mold condition. Within the new young slime mold the nucleus divides many times, and the mass becomes a gliding, feeding organism.

The cellular types lead a different existence, living in moist forest soil, and their fascinating behavior is best watched in a culture. Most familiar is the one called *Dictyostelium discoideum*. A culture dish containing nutrient agar is inoculated with cells of this species, and bacteria are added to act as their food. Their activities can then be watched through a microscope.

Each cell behaves individually for a while, creeping over the surface, engulfing bacteria, and reproducing by cell division. When the bacteria are nearly gone, a change comes over the population. One of the cells apparently gives a signal, and all those in its vicinity swarm to it. It is just as if an army sergeant called "fall in!" The signal is chemical rather than vocal, however, and is known to be a hormone called acrasin. What impels one cell out of many suddenly to secrete the hormone is still a mystery. Its influence is so strong that after bringing together a swarm of cells, it causes them to act as parts of a multicellular organism.

When food is plentiful the individual cells of the cellular slime mold wander and feed independently. When food becomes scarce one cell gives off a hormone signal that calls all the others to it, as shown in this photomicrograph. They then form multicellular organisms consisting of a stalk and a capsule that produces spores. The spores will be scattered by the wind, and under favorable circumstances each will germinate and become an independent cell that will wander and feed alone until the "call" comes again. (*J. T. Bonner*)

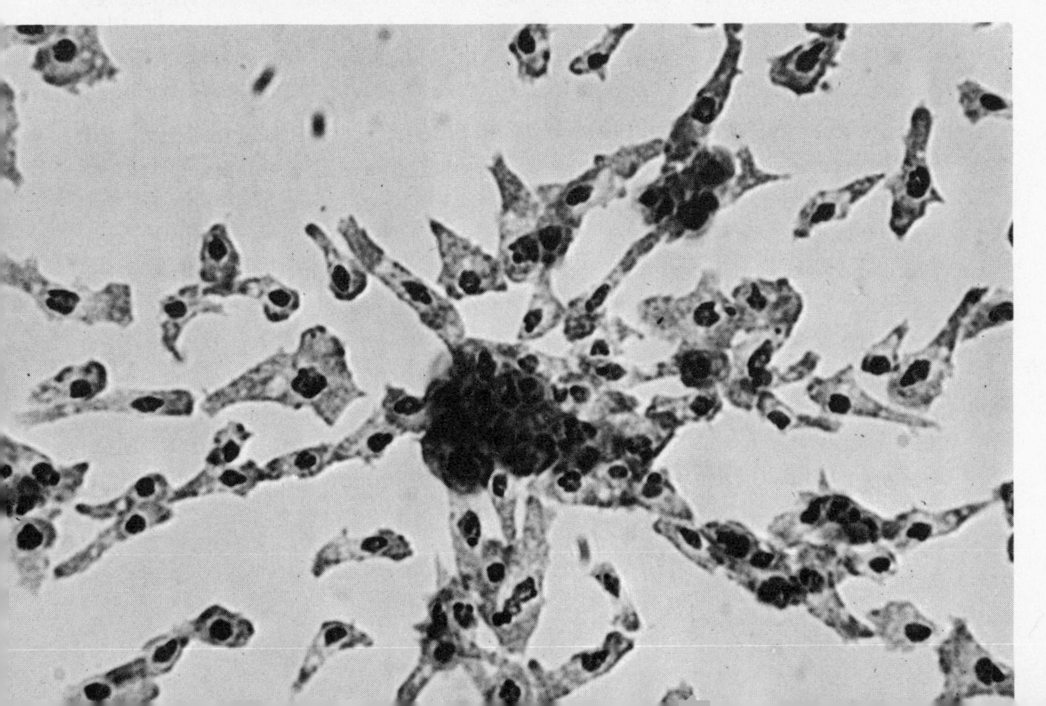

The swarm of cells becomes a sluglike creature about an eighth of an inch long, which creeps over the agar and is able to respond to such stimuli as light and temperature. In fact it is so sensitive to temperature that it can detect a difference of 0.0005 degrees. After wandering for a while, there evidently comes another signal. The front cells of the slug now rise up to form a stalk; the rear cells climb up the stalk to the top and form into a capsule in which spores form. In time, the spore case opens and the spores are liberated. In nature, they are carried away by the wind, and when conditions are favorable, each spore cracks open. An amoebalike cell emerges from it and once more begins the independent existence, wandering, feeding, and dividing alone until the "call" comes again.

The life cycle of *Dictyostelium* makes survival more certain. The aggregation of the cells into the sluglike form occurs when starvation threatens, and the spores they produce will be carried to new environments where food may be more abundant.

Dictyostelium discoideum illustrates how multicellular plants may have evolved from one-celled ones. In a one-celled organism one gene may have mutated, and this mutant gene may have directed the cell to produce a hormone which controlled the activities of neighboring cells, causing them to aggregate into a many-celled plant.

A different "teaming-up" process between algae and true fungi has resulted in the formation of lichens. Algal cells are wrapped in the mycelium of a fungus, and together they act as one plant. Most are formed with green algae, but some are formed with blue-green ones. Apparently the fungus provides a home and an anchor for the alga, protecting it from drought, intense light, and mechanical injury, while the alga makes food for both itself and the fungus. So many combinations of these basic components exist that 16,000 kinds of lichens are known, and they are so distinctive in shape, color, and texture that they are given species names. There are three general forms: crustose lichens, which are flat and grow close to the surface of rocks and the bark of trees; foliose ones, which have broad lobes and grow on rocks, trees, and the ground (even on desert sands); and fruticose ones, which are slender and branched and are often seen hanging from trees. Lichens range in color from grey and dull green to light green, pink, yellow, orange, and black.

They are often the first plants to appear on rock surfaces, where they begin the formation of soil and pave the way for plants of other kinds to follow. Mosses often accompany them or arrive shortly afterwards. The role of lichens in plant succession and the colonization of the earth is dramatic and of great importance. They have invaded all parts of the world, from the warm tropics to the Arctic and the Antarctic, and from sea level to the highest mountain tops. On rocky cliffs where no other plants dwell, you will see lichens, sometimes so widespread that you mistake their color for that of the rocks. The lichens grow most abundantly where rainwater washes over the stone, bringing not only moisture but nutrients washed out of bird and animal droppings. Some that have a high nitrogen requirement will grow only where they find such conditions. One kind in the high Rockies is even more selective: it will grow only where the cony or pika, the small rabbitlike "little chief hare" of Indian lore, marks its territory with drops of urine.

The lichens have a special mode of reproduction. They produce fragmentary masses formed into tiny balls called soredia that are rolled along by the wind. Where they are caught in cracks in rocks or in the bark of trees, they settle down and spread. The fungus member of the lichen team also produces its normal spores which are wind-borne, but these are ill-fated for it would be rare, indeed, that they would land among the necessary algal cells to reform the particular lichen. Whether a lichen is thus ever reconstituted in nature is not known. (Only recently have researchers been able to unite an alga and a fungus to produce a genuine, self-sustaining lichen.) The soredia are a more dependable means of reproduction.

Long ago, at times when food became scarce, people ground lichens into meal, from which they made bread. The "manna" of the Israelites may have been fragments or whole plants of foliose lichens carried by the wind over the deserts of Asia Minor. Lichens have been used to produce dyes, and though these have been largely replaced by synthetic ones today, litmus and orcein are still being made from them. The perfume industry uses some lichens in fragrances, particularly the "oak moss," *Evernia prunastri*.

Truly the fungi do not merit the term "lowly" which some people apply to them. For sheer variety of form and habit, for beauty and distinction, they hold their own with any other group of plants. They are the "other half" of the plant world, without which none of the green plants could survive, nor the animal life that lives with them.

chapter six

Mosses, Ferns, and Their Relatives

LIKE muted strings in an orchestra, the mosses and ferns play their subtle role in nature. Softly they cover moist ground; unobtrusively they seek a foothold in rock crevices, mellowing the contours of cliffs and walls even as they delight the eye with their intricate patterns of velvet and lace.

After life began in the oceans, the land remained empty and barren for a billion years, awaiting the time when plants could reach out to tap its riches and make them available to other—as yet unevolved—living creatures. The first plants to venture onto the shores must have been green algae that had evolved protective structures to allow them to push up onto drier land. Just what they looked like is lost to us, for the intermediate forms have not been found as fossils, but all land plants descended from them. Through one line came the mosses and liverworts, which did not evolve further and have not changed appreciably for 400 million years. Through another line came plants that developed first into the ferns and then went on evolving into what we know as the seed plants.

Meek though the mosses may seem compared to the more flamboyant species with whom they live, they are nevertheless venturesome explorers. They left the comfort of their original home on the shores, and through the ages became adapted to inhabit the whole earth, even places of cold temperature or little moisture. They now extend from the tropical and temperate zones to the tundra of the Arctic and the rocky islands of the Antarctic. They are found from sea level to elevations of more than 20,000 feet on Mt. Everest. Only in the oceans are mosses wholly absent.

In moist forests they cover the ground and the trunks and branches of trees. In the tropics, orchids and other epiphytes take root in the layer of humus they form. They cling to cliffs near waterfalls and cover the banks of streams and lakes, building soil for other plants to inhabit. Although most species prefer moist localities, some grow where drought is frequent and prolonged—among them those that grow on rock surfaces, where, along with lichens, they play a role in soil formation and help colonize barren areas. In dry weather they shrivel and turn brown, almost crisp, but they turn green and become plump again with rain. The Japanese, who are particularly adept at seeing possibilities in plant forms, make moss gardens in which they take advantage of even the slight differences in shapes and shades of color of the various kinds.

Mosses are relatively simple plants. They do not have specialized tissues to give them support and to conduct water and food. In higher plants the conducting tissues are accompanied by fibers that give strength to stems and allow them to grow to considerable heights. Mosses must remain small—rarely do they grow more than a few inches tall—and they gain support by growing in clumps, with the tiny plants holding each other up. An individual plant has a little "stem" that bears thin, spirally arranged "leaves." From the base of the stem, hairlike structures (not true roots) radiate into the soil to absorb water and minerals.

Although far removed from their original watery home, mosses still depend on water for reproduction, for they produce swimming sperms that must have water—even though only a film—through which to swim to the eggs. These sperms are formed in minute cases at the top of the stems of male plants, and eggs develop in tiny flask-shaped structures at the top of female ones. When the plants are covered with water from rain or dew, the sperms escape from their cases, swim to the eggs, and fertilize them. The embryo that results does not become separated from the parent plant as it would if the mosses bore seeds. Instead it remains attached to and nourished by it, and develops into a spore-bearing structure, a little threadlike stalk with a capsule at its tip. The little capsules are green at first, and as they ripen they turn red, brown, or orange, and contrast beautifully with the green beneath.

Each capsule has a pointed tip which falls off when the spores ripen, exposing an opening ringed with teeth. These teeth are often ornamented with cross bars or ridges, and are reddish in color. They are remarkably sensitive to changes in the humidity of the air. When the weather is damp, they bend inward and become coated with spores. When it is dry they swing outward, holding the spores up to be borne away by air currents. To be appreciated they should be viewed with a hand lens.

Although most mosses are land-dwelling, a few grow in or spread upon fresh water. Sphagnum, a soft, pale-green bog moss, may cover the entire surface of a pond or lake. As its dead remains accumulate, a mat is formed thick enough to support the weight of a man. Since there is water underneath and it quivers when walked upon, it is known as a "quaking bog." The peat—the mat formed by the sphagnum—may be harvested and used to improve soil or cut from old dried bogs to use as fuel. As time passes, various seed plants become established in the sphagnum—cranberries, cattails, and tamarack among them. Their roots add firmness to the floating layer. In time the pond becomes filled in completely, and in some areas trees eventually move in. Sphagnum also forms a ground cover in constantly wet rain forests —an incredibly soft, pale-green carpet.

Related to the mosses are the liverworts, flat-spreading plants with lobed parts, that grow in shaded places along stream banks or over stone walls or road banks in damp climates. Their name recalls the day when it was believed that plants shaped like human organs could be used to cure ailments of the part they resembled. This particular plant was thought to resemble the lobes of the liver—a resemblance that is obviously in the eye of the beholder, since in Ecuador it is called *mano de sapo,* "hand of the toad." The suffix "wort" is an old English word for herb or plant. One of the most widely distributed liverworts is *Marchantia.* This much-lobed, blue-green plant grows flat on the surface of soil or rock. At certain seasons it gives rise to vertical stalks that are terminated by disks that carry the plant's sex organs. Those that form sperms are tiny dimpled plates, while those that form the eggs are daisy-shaped. Both are as attractive as flowers. As in the mosses, the swimming sperms require water to get to the eggs.

On the other hand, the ferns and their relatives are better adapted to a land environment. The mosses and liverworts and the green algae from which they evolved remain fairly small because of their unsophisticated structure, but in the ferns, specialized conducting tissues make

Marchantia, a widely distributed liverwort. The male plants have discs on which sperms develop in tiny sunken cases. (*Alvin R. Grove*)

their appearance. The ferns and their relatives have true roots, stems, and leaves of considerable anatomical complexity. In the ferns and all the plants evolved from them, the food made by the leaves moves to all parts of the plant through a tissue called phloem. Water and minerals absorbed by the roots move into the tissue called xylem and are conducted by it to all parts. These conducting tissues and their associated fibers also give the plants support and enable them to attain great heights.

During the Carboniferous Period, 250 million years ago, ferns and their relatives—the giant club mosses and horsetails—were the dominant plants on earth and formed majestic forests that have not been surpassed in grandeur since then. They flourished in swampy areas and along the borders of inland seas. After death, their accumulated remains were prevented from complete decomposition by the surrounding stagnant water. As the inland seas advanced and retreated, successive layers of vegetable matter were laid down and in turn covered with sediments. During later geological eras they became buried under thick deposits of rock. The weight of the overlying rock, and the added pressure and heat as these layers were bent and folded by stresses within the earth, compressed them into coal. Depending on the amount of pressure, the coal became what we call soft or bituminous coal, or hard or anthracite coal. From soft-coal beds, intact tree trunks are sometimes exhumed.

Ferns and their relatives maintained their dominance until the Permian Period, when they were forced to share the earth with the newly evolving seed plants. The seed ferns, kinds that changed from the spore-bearing habit to produce seeds along the edges of their leaves, were the progenitors of the conifers and the flowering plants. No living examples of seed ferns exist today; we know them only as fossils. It is strange that the more primitive spore-bearing kinds should have outlived them, but simplicity is often the key to a species' survival.

Among the 8,000 species of ferns that exist today, only the tree ferns of tropical forests give an idea of what their giant ancestors were like. They tower sometimes to eighty feet, on tough, fibrous trunks, which are marked with triangles where the lower leaves have fallen, or retain the prickly black leaf bases as armor. Their almost rigid fronds spread in a circular roof from the top.

From these mammoth forms members of the fern family range on down to veritable miniatures; in substance they go from the tough bracken fern to the delicate maidenhair. The leaves of most are feathery and multiple-divided, but some are merely forked like a bird's foot or are quite plain and spatula-shaped. A few are shaped like hearts or four-leaf clovers, and their leaves are held on long slender "stems." Some others are long graceful ribbons. Wherever they grow they give a dreamy quality to the scene.

A bewildering variety of ferns can be seen in tropical moist forests, where they flourish on the ground, on fallen logs, and as epiphytes on trees among the orchids and bromeliads. Some ferns are as tall as a man, near-giants in their own right although they can't rival the tree ferns, and beneath them grow myriad other sizes and shapes. Among the epiphytes overhead are tiny ferns whose little leaves march in single file along the branches. On the tree trunks cling the massive staghorn ferns, whose broad antlerlike leaves spread from a cushion base.

Ferns of many kinds also inhabit the moist forests of temperate climates and occur in open fields and on cliffs and rocky hillsides. It is rather surprising to find them on glacial boulders, until one realizes that there they find some soil prepared for them by the pioneer mosses and lichens and that they find a supply of water in cracks in the rocks. Here they, too, play a role in plant succession, helping to build more soil for the seed plants that will come in later. There are even a few kinds of fern that grow in water or float upon it like giant duckweeds.

All of the land ferns reproduce by spores that develop in spore cases on the underside of the leaves or on structures made from modified leaves. At certain seasons, you will see rust-colored spots. Each spot, called a sorus, consists of hundreds of minute cases, in each of which many spores develop, often 64 to a case. A few ferns, including the staghorn, have the spore cases spread in a wide patch. When the spores are ripe, the cases open, the spores are catapulted into the air, and the wind carries them to new locations. If you grow ferns in a greenhouse, sometimes you will find neighboring plants covered with the fine dust of the spores. If you pass your hand gently under a leaf, or shake it over a piece of paper, you can gather the spores.

The life cycle of ferns consists of two kinds of plants, an alternation of generations. The spore germinates and develops into a little flat heart-shaped plant that looks somewhat like a small liverwort. It is called a gametophyte because this is the stage that produces the gametes, or reproductive cells. The sperms and eggs form in tiny structures on the undersurface of the heart-shaped plant. As in the mosses, the sperms are of the swimming type and must have water to reach the eggs. The fertilized egg develops into the familiar sporophyte stage that we know as the fern plant—at first attached to the gametophyte, but

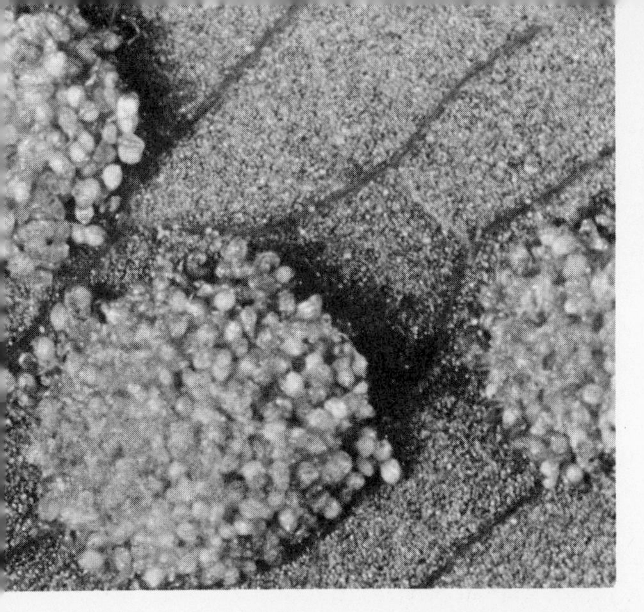

Ferns have clusters of spore cases on the underside of the leaves. Each tiny case holds a great many spores, and when they are ripe the cases pop open and propel the spores into the air.
(*Photo by Tom Northen*)

The bracken fern adds beauty to the forest floor.
(*U.S. Forest Service. Photo by Leland J. Prater*)

becoming independent as the tiny parent plant dies. The first leaves are often quite different from those of the mature plant, being more simple, usually without the leaflets or leaf divisions that will appear on later leaves. The stem of the fern is a rhizome, or ground stem, often covered with fuzzy hairs, and from it grow true roots. Buds along the rhizome give rise to the leaves or fronds.

The water fern *Marsilea* has developed a special way to reproduce. It grows rooted in the mud along the shallow edges of ponds, its four-leaf-clover leaves floating on the surface. It produces bean-shaped structures near the base of the stems in which male and female spores develop separately. These are discharged into the water in a gelatinous mass. The male spores develop into tiny sperm-producing gametophytes, and the female spores into gametophytes that hold one egg cell each. The sperms escape, swim to the eggs and fertilize them, and each fertilized egg develops into a new fern plant.

Relatives of the ferns include several plants that most of us know and love, among them horsetails, selaginellas, and lycopodiums. They evolved from the same ancestors as the ferns, but like the mosses, they were evolutionary ends in themselves.

Modern horsetails (species of *Equisetum*) are found along streams or on sandy banks. Pioneers called them scouring rushes and used them to scour metal utensils, a use for which their silica content lent itself quite well. They have jointed stems that you can pull apart and fit together again by means of cup-shaped bracts. At the top of the stem is a conelike structure in which spores are formed. Under a microscope it can be seen that each spore has ribbons attached to it. These coil around the spore in damp weather and uncoil when it is dry, pushing the spore out of the case and giving it "wings," which aid its transport by wind. As in the ferns, the spores grow into inconspicuous

This beautiful lycopodium is called the "Queensland tassel fern" in its native Australia. It also occurs in New Zealand and Fiji. (*Rebecca T. Northen*)

little flat plants that form eggs and swimming sperms, and the fertilized egg grows into the horsetail plant.

The club mosses are well known for the unique beauty and peculiar habits of some of their members. They embrace both *Selaginella* and *Lycopodium,* and almost everyone knows one or more of each, although their names are often confused in florist shops and garden stores where they can be bought. Both have tiny waxy leaves. In *Selaginella* the leaves are scalelike and lie flat to the stem, two rows of small ones alternating with two rows of larger ones. In *Lycopodium* the leaves are all the same size and are arranged spirally around the stem, sometimes lying fairly close to it but often sticking out at an angle. Both are much branched. The sellaginellas and some of the lycopodiums produce their spores in "cones," cylindrical structures in which the "scales" are modified leaves. The cones can be insignificant in size or quite showy and decorative.

There are five hundred species of *Selaginella.* Almost everyone has seen the resurrection plant. It can be bought in the dime store as a brown, lifeless-looking mass. When you put it in water its feathery fronds uncurl and become soft and green. Then there is the "peacock fern," so delicate that it can be grown only where it is warm and very damp. Its beautiful fernlike fronds are extremely lacy and shine with a purple iridescence.

The "ground pines" found in woodsy places are some of the one hundred species of *Lycopodium.* Their trailing or upright stems, tipped with variously shaped cones, are up to a foot long and add greatly to fall bouquets and Christmas decorations. In Central America emerald green lycopodiums grow as epiphytes in damp forests. Before gathering them you must be sure they are not harboring a little poisonous green snake, which is about the same thickness as their stems. Perhaps the most beautiful of all lycopodiums is the epiphytic "Queensland tassel fern" (note the many mistaken terms used for plants in this group), native to the Old World tropics but nicknamed for Queensland, Australia. Its copiously branched stems, set with waxy, triangular leaves jutting out at right angles, are tipped with veritable tassels of long, delicate cones.

Nature was surely in a gentle mood when she created the mosses and ferns and their relatives. She also gave some of them hard jobs to do but made them tough enough to perform them. This group of plants exemplifies Nature's requirement that each living thing play its role in its own niche.

chapter seven

The Ancient and Honorable Order of Conifers

THE invention of the seed and the mechanism that produces it was one of the most significant innovations in the history of the world, even more significant than man's invention of the wheel. Before seed plants came into being, the world's land plants reproduced by spores alone, and although it was an extravagant and precarious method of reproduction, no one can say it wasn't effective—by this means they colonized the earth. But as with many inventions, while the old still works, the new has many advantages. This is particularly true of seeds.

A great deal of parental care is lavished on the seed. The young plant it contains is as carefully outfitted for life as a human youngster preparing to leave home. Protected and nourished by the parent, the embryonic plant is brought to a certain stage of development, surrounded with a storehouse of richly concentrated food, and covered with a weatherproof coat within which it can usually survive drought, heat, and cold. The embryo is dormant, but is prepared to enter into active growth as soon as conditions are right. Embodied in the seed is a signal that is triggered only when conditions are favorable for the growth of the new plant. When the seed germinates, the embryo quickly becomes active, drawing on the stored food until its own roots and leaves are sufficiently developed to support it. The "child" is then launched into independent life.

The invention of the seed took place about 300,000,000 years ago, long before the appearance of the dinosaurs. Among the giant spore-bearing plants that covered the earth, such as horsetails, club mosses, and ferns, there evolved a kind of fern that produced seeds along the edges of its leaves. From the seed ferns there developed some primitive cone-bearing species, among them the cycads and the ginkgo, and then the fully developed conifers. All of these outlived their common ancestor, the seed fern, and many remained to outlive the dinosaurs that shared the earth with them for additional millions of years. The cycads and ginkgo (a single species, *Ginkgo biloba*) survive in smaller numbers now, but the conifers became the dominant plants on earth until the more highly specialized flowering plants came along about 160,000,000 years ago.

Leaves and seed of the ginkgo, a tree of very ancient lineage. (*U.S. Forest Service. Photo by W. D. Brush*)

The conifers and the flowering plants produce their seeds in quite different ways and are given names that characterize their habits. The conifers are called gymnosperms, a word meaning "naked seeds," from the Greek *gymnos,* naked. (Our word "gymnasium" is a reflection of the fact that Greek athletes wore no clothes during their contests.) The seeds of the conifers are not enclosed in a pod or fruit, but instead lie on little shelves in a cone, from which they fall free when the cone ripens. The flowering plants are called angiosperms, from the Greek *angeion,* a case, vessel, or capsule; and their seeds are contained in a dry pod or a fleshy fruit.

The mechanism that set plants free from a watery environment was pollen. The pollen grain contains the male reproductive cells, the sperms that fertilize the eggs. But instead of having to swim to their destination as they once had to do, they are reduced to mere nuclei and are delivered directly to the egg cells through a conducting tube that grows from the pollen grain to the egg. This frees the plants' reproductive cycle from dependence on water, and allows them to live in a land-based environment.

The gymnosperms do not have flowers—there are no sepals, no colorful petals, no nectaries. Instead their reproductive organs consist of male and female cones, both very small when they first appear in the spring. The beautiful little male cones are as pretty as flowers; however, they are short-lived and often escape notice. They are borne in clusters in early summer, usually on the lower branches, and are purple, pink, or yellow, depending on the species. Soon after they appear, they open and shed an enormous amount of pollen that drifts away in the wind. Often the air is tinted with it, and it covers everything with gold dust—other plants, picnic tables, parked cars. Only the grains that happen to land on the female cones ever have a chance to function; the rest merely exemplify nature's extravagant overproduction in order to make sure that the law of averages will assure propagation of the species.

The young female cones are tiny, often purple, consisting of delicate scales arranged spirally on a central stem. On the upper surface of each scale are two potential seeds, called ovules. The air-borne pollen dusts over the female cone and sifts to the bottom of the invitingly open scales. There, the pollen grains come in contact with the ovules. Each pollen grain sends out a tube that carries a sperm to the egg contained in the ovule, fertilizing it and starting a seed on its way to development.

The tiny male cones of the jack pine, ready to shed their pollen. (*U.S. Forest Service. Photo by W. D. Brush*)

After the eggs have been fertilized, the cone swells shut, enclosing the developing seeds. It grows in size until it is large, fat, green, and succulent looking, and remains that way until the seeds are ripe. When their development is finished—that is, when the embryo with its supply of food is ready and enclosed in its tough seed coat—the cone becomes dry and woody and the scales separate again to liberate the seeds. In spruces and firs the cones complete their development in one season, but in many pines the seeds do not mature nor the cone open until the second year. Occasionally some cones remain stubbornly shut even after the seeds within are ripe. In lodgepole pine and jack pine, for example, some cones stay closed until a fire sweeps through the area. The heat brings about the cones' opening, and the liberated seeds then reforest the area. It is as if Nature saves a few caches of seed so that they will be on hand in the case of just such an emergency.

The cycad and the ginkgo show the half-way step between the ancient ferns and the true conifers in a rather amusing way. The plants themselves are separately male and female. The pollen is carried to the fe-

The female cones, small and inconspicuous at this stage, will receive the pollen and then swell shut. During the following months they will enlarge greatly, and finally will open to shed the mature seed.
(*U.S. Forest Service. Photo by W. D. Brush*)

male by the wind, as in the conifers; after it lands on the female organ it dutifully puts out a tube to conduct the sperm to the egg. But within this tube, instead of the mere nucleus one might expect to find in seed plants, is a pair of huge swimming sperms. They are even covered with the cilia or flagella, the little taillike organs that beat the water like oars, which they inherit from their fern ancestors. Thus representative of an earlier evolutionary stage, the swimming sperms are held captive in the pollen tube of later development and are delivered to the egg cells without ever being turned loose to use their swimming ability.

The cycad is a beautifully decorative plant grown in greenhouses and tropical areas. Its palmlike leaves grow in a circular crown in the center of which is a single cone, the male on one plant, the female on another.

The ginkgo would have become extinct, it is thought, if through long ages it had not been revered in China and Japan and kept flourishing in temple gardens. It is called the maidenhair tree, and looks nothing like its conifer relatives. The leaves are flat bilobed fans (hence the species name, *biloba*), which are shed in the fall. The female cone would not be recognized as such, for it is a round structure about the size of a cherry, containing a single seed. While the seed is "naked," as in the conifers, the seed coat is a fleshy skin with an unpleasant odor. Moreover, when the seeds fall, their mushy coats make the pavement slippery. The ginkgo, never exposed to anything like our city life in its native haunts, has become a valuable city tree in the United States. Strangely, it can tolerate the dust, heat, and fumes far better than some of our native trees, and it is being imported to lend its green beauty to places where other, more advanced species are being choked to death. Obviously, the male tree is chosen for planting, unless the purpose is to raise young ginkgos from seed.

The conifers are all trees or shrubs. They have either small, scalelike leaves, as are found in junipers, cedars, and *Sequoiadendron,* the "Big Tree," or needlelike leaves such as those of pine, spruce, larch, fir, redwood, hemlock, and Douglas fir. In pines the needles are borne in clusters of two to five, in larch in clusters of more than five; in the others they are borne singly. Those of spruce are square in cross section and sharp pointed—so sharp that a spruce does not invite you to sit under it for a picnic or to use its branches to make a camping bed. In contrast, the needles of fir, hemlock, and Douglas fir are blunt and flat, and feel soft to the touch. Most conifers are evergreen; among the exceptions are the larch, the bald cypress, and *Metasequoia,* which shed their needles each autumn.

Several of the conifers are making contributions to lands far from their native habitats. The piñon pine of our western mountains has been transplanted to Lebanon, where the famous Lebanon cedars have almost vanished. There the piñon flourishes, and the hills above Beirut are well wooded with it. It is producing a new trade product—its edible pine nuts—and is furnishing lumber and fuel wood. The Deodar cedars from the Himalayas now flourish in parks and along streets in Los Angeles.

The little Monterey pine (*Pinus radiata*) is native to a mere eight miles of dry hills on the Pacific Coast, and grows so slowly there that it has remained forever small. It grows more rapidly when moved to areas of greater rainfall. Transplanted to the damp climates of New Zealand, Australia, and Chile it has become the fastest-growing timber tree in the world, increasing in height as much as eight feet a year. It makes its most rapid growth while young, but reaches almost a hundred feet in 24 years. Sawmills and pulp factories are springing up in these countries to make products for both local consumption and for export, and to add greatly to the national income. It has been a special boon in New Zealand, where the huge native trees, such as the Totara, from which the Maoris fashioned their tremendous canoes, and the Kauri, the gigantic gum tree, were decimated by European settlers. The native trees are too slow growing to fill the needs of the expanding population. Actually, the Monterey pine is thought to have made its way to New Zealand by way of Australia, through seeds carried by prospectors.

Another tree from the California shore, the Monterey cypress (*Cupressus macrocarpa*), is spreading over New Zealand. Almost everywhere you look long ranks of its dense forms stretch between fields and across hills in the form of shelter belts, their dark green contrasting with the emerald green of the grass. In its native home the Monterey cypress is twisted and dwarfed by the salt winds of the ocean. But transplanted to the constantly humid and more gentle conditions of New Zealand it becomes a fat, rounded pyramid.

In the United States we are not only tree lovers but tree users. We use wood lavishly, especially that of conifers such as the southern pines, Douglas fir, ponderosa pine, cedar, hemlock, spruce, fir, and others. Wood from such trees goes to make lumber, ties, posts and poles, and wood pulp for paper and rayon. We use more paper than all the rest of the world combined, over 400 pounds per capita annually. It behooves us to protect our forests and replant land that has been denuded.

In our 48 contiguous states there are about 650,000,000 acres of forest land, populated mostly by conifers. The conifers frame the grandeur of many national parks and monuments. Throughout the ages man has sought the beauty and solitude of forests for inspiration. Surely no one, no matter how burdened, could walk through the redwood forests of California, or the spruce and cedar forest of Olympic National Park, or the forests of the Great Smokies, without feeling his spirit lifted. There is something about the serenity of spirelike conifers that brings inspiration to man.

chapter eight

Flowering Plants

THE flowering plants are the most recent additions to the earth's flora. From the time they began to evolve, 160,000,000 years ago, they have graced the land and its waters with color and perfume and have nourished its inhabitants, providing man with most of his food and fibers and feeding a host of wild animals from insects to elephants. Their diversity allows one or more species to thrive in practically all habitats, from desert to lake to forest, from the tropics to within the Arctic Circle, and from sea level to alpine summits. They come in all sizes, from some that are a quarter of an inch tall with flowers the size of the head of a pin, to larger ones with dinner plate flowers, and on to our hardwood trees.

Although for the plants, and for animal life as well, the flowers are most important as the prerequisites for fruits and seeds, we enjoy many of them for their flowers alone. We nurture them in our gardens and homes and give them as tokens of affection or sympathy. Often it is a plant that typifies a place for us—the magnificence of the flowering cherries in Washington; an expanse of pink blossoms in an orchard on a back country lane; the roof-high poinsettias of our South; the tulip fields of Holland; the sweetness of crisp white frangipani in Hawaii or around temple gardens in Ceylon; the voluptuous violet-blue Jacaranda in Mexico; an orchid clinging to a branch in a tropical forest. The

The flower of the tulip tree, *Liriodendron tulipifera,* is a primitive type of flower, with three sepals, six petals in two rows, and numerous stamens and pistils. (*U.S. Forest Service. Photo by W. D. Brush*)

fragrance of plants is nostalgic. Even though in our adult years we may cultivate more sophisticated tastes, the perfumes from the gardens of our childhood live with us, and all the memories come flooding back at one whiff of a familiar scent.

Most of those who cherish flowers have never really seen their most appealing and intriguing parts. Those who have the patience and curiosity to look at them with a hand lens or to photograph them at close range will enjoy a thrilling new world in the shapes and color patterns, particularly of the reproductive parts. No mathematician or statistician could outdo nature in creating such incredible and bewildering variations on the basic theme. Our appreciation of flowers and of the marvels of reproduction is enhanced by understanding that basic pattern and the events that take place in their life cycle.

A complete flower has four whorls of parts (we say a *complete* flower because some are incomplete). The outer whorl is composed of the sepals, usually green and of leafy structure, which cover the initial flower bud and fold back when the flower opens. Within the sepals is the next whorl, the petals, often but not always beautifully colored, waxy, satiny, or velvety. Within the petals comes the third whorl, the male reproductive organs or stamens, each consisting of a vessicle—the anther—that produces pollen, held up on a filament, or delicate stalk. In the center of the flower is the seed-forming or female organ, the pistil, resembling a vase in shape. Its swollen base is the ovary and holds the potential seeds or ovules; it is the ovary that becomes the fruit. The slender neck of the vase is the style and the top is the stigma. The top is not open as in a vase but is closed and has a sticky surface that receives the pollen. The reproductive organs are the essential parts of the flower. The petals are pure decoration.

For every seed—in other words for every embryo plant—a sperm cell brought by a pollen grain must fuse with an egg cell in an ovule. Both the sperm and egg cells contain half the number of chromosomes of their parents, so that in fusing they reconstitute the full number for the species. The principle is the same for plants, animals, and human beings. The characteristics of an individual are determined by both the male and female parents; it receives half of its genetic material from each.

Diagram of parts of a flower. (*Drawn by Abbie M. Current*)

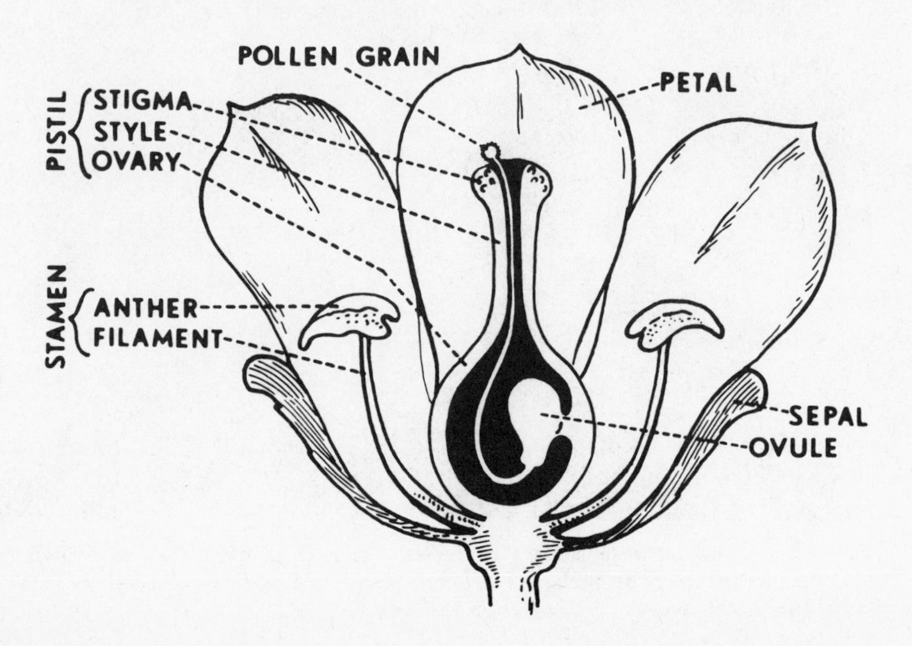

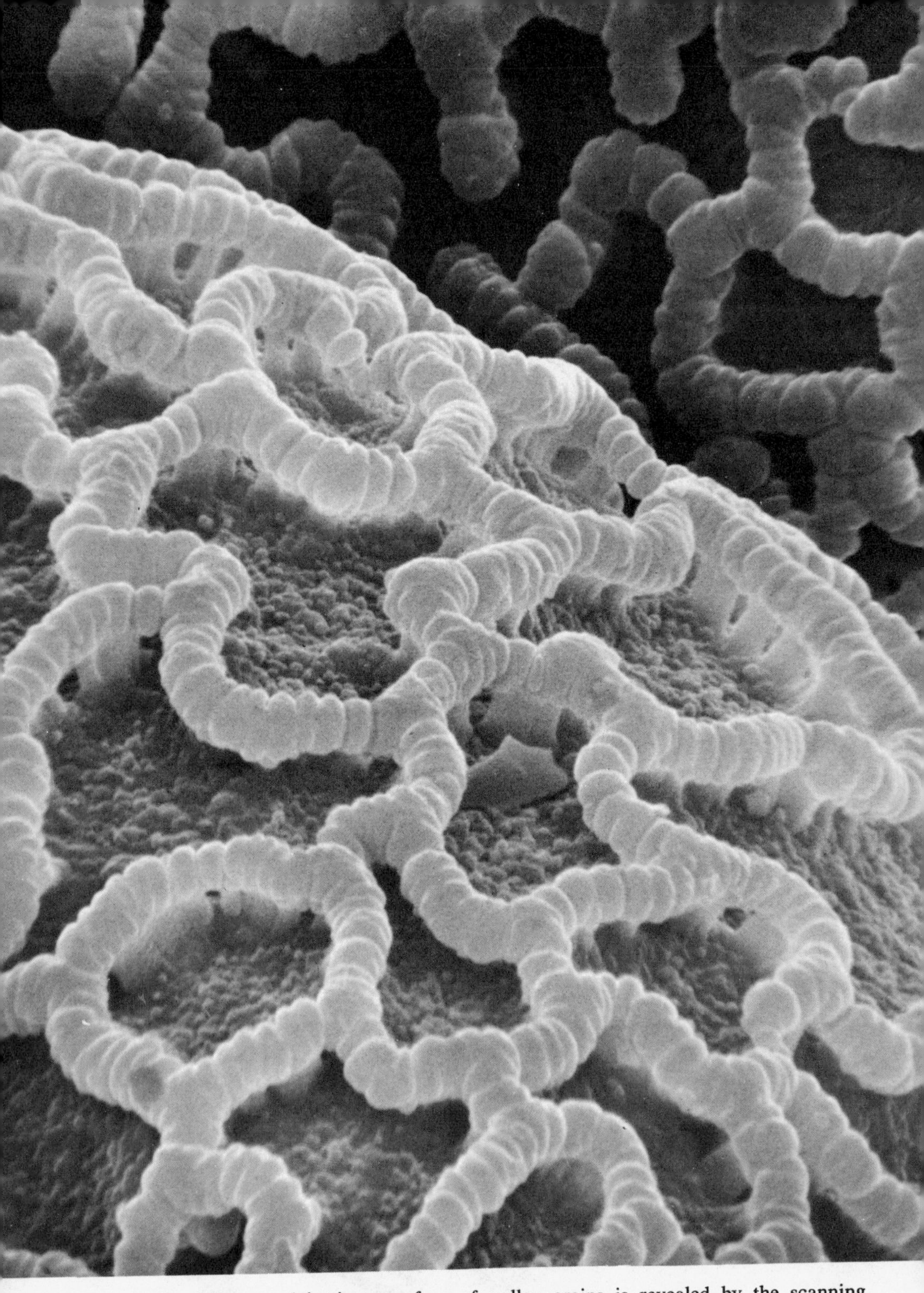

The delicate and intricate surface of pollen grains is revealed by the scanning electron microscope. The pollen of *Lilium longifolium*, specially treated to allow the ridges to be seen. (*J. Heslop-Harrison*)

Within the anther certain cells undergo reduction division, which halves their number of chromosomes, carriers of the genes. The resulting cells are immature pollen grains, each with one nucleus. The nucleus then divides, giving a pollen grain with two nuclei.

While the pollen is being formed, events are taking place in the ovary where one or more ovules is developing. A particular cell in each ovule undergoes halving of its number of chromosomes, and during the process four cells are formed. Three of the four come to naught, but the fourth goes on to become an embryo sac containing an egg.

The next step is one of logistics, of getting the pollen to the stigma. Some species have dusty pollen that is carried by the wind; others float theirs on water. Most kinds have sticky pollen that adheres to the anatomy of a visiting insect, bird, or bat and is thus carried to the stigma of a nearby flower. Pollen grains of each species differ from those of all others. They vary in size, color, and shape, and may be plain or sculptured with crests, ridges, or knobs.

Each pollen grain perched on the stigma sprouts a tube that grows downward through the tissues of the style and into the ovary, where it seeks out an ovule. While it is growing in length, one nucleus divides to form two sperms. When the tube makes contact with an ovule, it penetrates its outer tissues and releases the two sperms into the embryo sac. One of the sperms fertilizes the egg, which then develops into an embryo.

Fertilization of the egg triggers the whole process of seed and fruit formation. The ovule itself, containing the embryo plant, becomes the seed, replete with food to nourish the embryo when it begins its new life, and a tough outer coat to protect it until that time. Hormones produced by the developing seed move into the ovary and stimulate it to develop into a fruit. In most plants the ovary is tiny to start with, but as seeds develop it grows enormously. The one-eighth-inch ovary of a peach develops into a fruit eight inches in circumference. From an almost equally small ovary the watermelon grows into a fruit weighing ten to twenty pounds. Some fruits contain one seed; some have hundreds of thousands. Some seeds themselves are large, an avocado's, for instance; some tiny, as in a poppy. Some fruits are fleshy, such as the tomato, and others hard like nuts. They fall into various categories, but botanically, all structures that develop from an ovary and contain seeds are termed fruits—including such varied types as olives, cherries, cucumbers, pea pods, and coconuts.

The pollen of *Cosmos bipinnatus*. (*J. Heslop-Harrison*)

The fact that a pollen tube finds its mark in an ovary with one ovule gives no cause for wonder. Nor is it surprising that in an ovary with ten or a dozen ovules, all should become fertilized—especially since we know that many pollen grains were deposited on the stigma. When you empty into your hand the almost uncountable number of seeds from the little salt-shaker capsule of a poppy, you might pause to consider how a pollen tube found each one. Here were hundreds of ovules in the ovary. Thousands of pollen grains had to be deposited on the stigma, each to send its tube into the ovary, and most to find an ovule not already found by a competitor. Does the tip of the pollen tube as it approaches an ovule sense that this one is already fertilized, and does it then pass on to the next in line? How does a pollen tube find the last unfertilized ovule among the tremendous numbers? These are mysteries as yet unsolved.

Yet the number of seeds in a poppy capsule is nothing compared to some others in the plant world. Orchids have hundreds of thousands, even millions of seeds in one pod—extremely tiny ones, of course. Those of the green swan orchid were patiently counted and proved to number 3,770,000. How did a pollen tube find the last one after the 3,769,999th ovule had been fertilized? The orchid does not have numerous anthers as do other plants. Its reproductive organs are fused into one structure, with the single anther at the tip and the stigma below it. The pollen grains are bound into waxy pellets about the size of the head of a pin, called pollinia. An insect may transfer one or several pollinia to the stigma of a flower. Just think of the numbers of pollen grains that have to be contained in each of those tiny pellets!

Animals have elaborate techniques for securing mates, often involving battles or fancy displays and courting dances. Plants are more passive—they have to wait for pollen to be brought to them. But they are selective, in that only genetically compatible pollen can function. This means that the fertilizing pollen must be from at least a closely related species; hybrids occur or can be made among such groups of species. If an insect brings pollen from a daisy or delphinium to a rose, which are not closely related, the pollen just sits on the stigma and either cannot sprout, or if it does send its tube into the ovary, its sperms cannot fertilize the eggs. Species that *could* interbreed are often prevented from doing so in nature because of barriers of one sort or another, as we shall see later, but the majority will accept pollen only from their own kind. Some can be self-pollinated—that is, a flower can be pollinated with its own pollen or that of other flowers on the same plant. Most, however, reject their own pollen, requiring pollen from other plants of the same species. This cross-pollination encourages crossbreeding, with more varied and sturdier offspring.

To the dismay of some amateur orchardists, many fruit trees are quite selective about their mates, among them certain varieties of apples, plums, and pears, and all varieties of sweet cherries. Perhaps the most popular sweet cherry is the Bing, which rejects its own pollen and also that of Lambert and Napoleon. An orchard containing all Bing trees or a mixture of these three will never produce a single cherry! But Bing doesn't reject all suitors; it welcomes Black Tartarian and Black Republican. Planting a few trees of either of these along with Bing trees insures a good crop of fruit. The secret is that all of these are selected strains of the same species. Evidently some strains are so much alike genetically that they refuse each other's pollen as surely as they do their own, and accept the pollen of those that are just enough different to function as cross-pollinators. The characteristics of the fruit, of course, are determined by the female parent—the fruit is formed from the ovary—while the male parent contributes to the genetic content of the embryo.

Many plants go even further to assure cross-pollination. For instance, some flowers' anthers mature before the stigmas become receptive. In hollyhock, for example, the cluster of stigmas remains hidden and protected down within the anthers until the anthers have opened and shed their pollen. Then the stigmas grow taller and spread out to receive pollen from other flowers. Orchids are built so that visitors receive an application of pollen (by means of glue or a sticky disc) as they leave the flower rather than as they arrive.

In mammals, the division of a species into two separate sexes insures crossbreeding. Some plants have evolved separate sexes, too, with male and female flowers on different plants. Male hollies, cottonwoods, willows, and a number of others produce the pollen, and their female counterparts the fruits. If a homeowner wants a holly tree loaded with berries in winter, he must be sure not only to plant a female but to see that its flowers are pollinated. He could plant a male tree nearby, or he may be able to obtain the services of a "stud" at flowering time. A few branches of a male tree with ripe pollen, maintained in water and placed among the branches of a female tree, will insure pollination and berry formation. To make things vastly easier, male branches can be grafted onto female trees to function permanently—and this can be done with cherry trees as well.

If you have something to sell, you have to announce the fact to prospective consumers. So it is with plants that rely on members of the animal kingdom as pollinators. To gain their services, what better method than to offer them food and drink, and to let them know by eye-catching displays and delightful odors that the feast is ready? Some of their schemes are straightforward and honest, but others can only be called underhanded, even nefarious. Most plants offer food most generously, even providing their visitors with enough to feed their young and the rest of the household. However, some attract their customers by appealing to their senses or by offering special services; others, after promising a reward, trick them into performing pollination and send them away empty-handed. The fascinating variety among flowers—their size, shape, colors, odors, and textures—is explained by the variety of agents they rely upon.

Among pollinators, bees are the most numerous. The flowers that attract them are generally pleasantly fragrant and attractively colored, the ones that appeal to us as well. Similarly, flowers that are odorless to them are odorless to us. They are attracted by shades of yellow, blue, pink, white, purple, and to a lesser degree, red. They actually cannot see red (for it appears black to them) but they can certainly see the mass of yellow anthers within a red flower. Also, since their eyes are sensitive to ultraviolet light, they may see some colors and patterns invisible to us. To accommodate bees, many flowers offer them broadly spread petals as landing platforms, and some even mark the way to the nectaries by stripes or locate them with bright spots.

Bees get a double reward from their chosen flowers, for in addition to nectar they take loads of pollen home to feed their young. As they wallow (there's no other word for it) among the anthers, their fuzzy bodies become covered with the delectable globules and they have to pause now and then to scrape them off. They push the pollen into little baskets formed by bristles on their hind legs. Watch bees at work some time and you will see some with the packed yellow masses near their knee joints. Their alimentary tract has a specially enlarged pouch, or honey crop, in which they carry nectar. They really transport quite a weight by the time they are ready to return to the hive.

On a trip from hive to flowers and back again, bees usually visit only one kind of flower. Sometimes they concentrate on one kind for days. As a result, very little pollen is wasted; what they deposit on the flowers is of their own kind and therefore able to function.

Butterflies and moths are busy pollinators, too, and the nectar they sip with their long proboscises is usually the only food they eat in their adult stage. They visit many of the same flowers bees do, but some flowers are especially constructed for them, with deep nectaries such as those of columbine and lychnis. Butterflies are not interested in pollen, nor are most moths; but they transfer it inadvertently. Night-flying moths are accommodated by flowers that either open at night or stay open for some part of it, and that are usually white, creamy, or light pink so that they can be seen. They are strongly fragrant. Some flowers that depend on pollination during the dark hours emit their fragrance only when their pollinators are abroad—some just at twilight, some during the night, and others just before dawn. Curiously, these kinds may be open during the day, too, but are not fragrant at that time. The "lady of the night" orchid (*Brassavola nodosa*) is one example: in fact, its blooms stay open for two or three weeks, and each evening it sends out its perfume to lure a particular species of moth.

Hummingbirds are brilliant little dynamos that dart and hover, flickering in and out of flowers in an instant. Their hum comes from their wings' beating seventy times a second. Their tenth-of-an-ounce weight allows them to perch on a stem of grass, they weave spiderwebs into their nests, and yet they are so pugnacious that they drive off birds a hundred times their size. The Ruby Throat, the only species found east of the Mississippi River and the most common in the United States, migrates as much as 2,000 miles each spring and fall. It is known to fly directly across the Gulf of Mexico, a trip that necessitates 500 miles of continuous wing action.

These little jewels of the bird world need enormous amounts of food to power their furious activity. Flowers that cater to them furnish copious amounts of nectar, which they draw up through a tubular tongue. That slender, flexible tongue is also quick to pick up any small insects on the flower or in nearby spiderwebs, thus adding protein to the hummer's diet.

Hummingbirds live only in the Western Hemisphere, from Sitka to Tierra del Fuego, with 8 kinds in our West, 55 in Mexico, and 163 in the Ecuadorian Andes. While they live in all kinds of environment from desert to mountains, they are concentrated in the high Andes. This is not surprising when you know that butterflies and bees cannot survive in the highest elevations; therefore the hummingbirds have the flowers to themselves. And flowers there are aplenty, orchids and gesneriads especially. They have access to thousands of kinds at lower elevations as well, albeit with increasing competition.

Hummingbirds are not limited to flowers with tubular nectaries, although these are especially appealing. They can obtain nectar from many kinds that attract bees, and they are known to follow bees to find their sources of nectar. However, flowers that are especially constructed to attract birds do have deep nectaries and provide guides for their beaks—narrow tubes, rigid channels, or rows of firm protuberances. Many such flowers omit the expanded petals, since hummingbirds do not need a place to land. Some flowers are in the form of vases, upright and filled to overflowing with nectar, or hanging downward with great drops of nectar inside.

In pushing into a flower, a bird forces its head against the anthers and picks up pollen on its feathers, brushing it off on the stigma of the next flower in the same way. Some flowers hold their anthers and stigma arched over the entrance to the nectary in just the right position to achieve this. The Indian paintbrush places its pollen on the top of a hummingbird's head, and a species of honeysuckle places it on the feathers at the corners of the hummingbird's mouth. Insects visiting these same flowers may miss the pollen entirely, although they may be perfectly able to get the nectar. The *Marcgravia* of Central America has a wheellike arrangement of little suspended slippers, and in hovering over them hummingbirds effect pollination with the tips of their wings.

Other birds pollinate flowers, too—the brush-tongued parrots (also called "honey eaters") of Australia and the Pacific islands, and the sun birds of Africa and India. Even in this country, birds you would not suspect carry pollen regularly or occasionally—the Saguaro cactus, for instance, is pollinated by a dove. All birds are attracted by flowers of brilliant colors, and striking contrasts of orange and blue, scarlet and green, or red and yellow, such as those found in the bird of paradise, bromeliads, heliconia, and fuchsia. Of all colors, red is the avarian favorite, and in the tropics, trees with red flowers are often crowded with birds vying for nectar. After red, the color preference (in descending order) is pink, orange, blue, yellow, white, green, and maroon.

A dove pollinating a flower of the saguaro cactus. (*U.S. Dept. of Agriculture*)

When a plant relies on a single species to be its pollinator, no other species can perform the job. In thus limiting itself, the plant must make itself so attractive to that species as to be irresistible. It must look and smell exactly right or the pollinator won't find it. This is virtually a matter of life or death for the plant and often for the pollinator. By means of shape, nectar, color, odor, or by means of sometimes elaborate tricks or devices, it must inveigle, detain, and dispatch its chosen servant.

Members of the genus *Yucca* depend absolutely on the services of the little white, three-eighths-inch pronuba moth; no other species will do. Neither can the moth survive without the yucca, whose seeds are the only food its larvae can eat. The moth winters in the ground as a larva, and emerges and mates just before the yucca comes into bloom. The female flies to a flower, gathers some pollen which it rolls into a little ball under her chin, and then heads for a different flower. There she punctures the ovary with her ovipositor and lays several eggs within it. Before leaving—and this is the most remarkable act in the sequence—she stuffs the ball of pollen into the tip of the stigma, which is cup-shaped to receive it. By pollinating the flower, the moth starts the development of the seeds upon which its larvae must feed. There are always more seeds formed than the larvae can consume, leaving some to propagate the yucca.

Flies are a group of insects that human beings find anything but aristocratic. But just as human beings in various lands have different ideas of what constitutes beauty or "class," the plants have their choices also. The creature that makes a plant's survival possible must surely be the most desirable of all to that plant.

Some species of flies like sweet scents and pretty colors, and the flowers that appeal to them oblige accordingly. One of these, the beefly, which looks quite like a bee and has a long sucking proboscis, is the pollinator of the primrose. But the flowers that appeal to bluebottle and carrion flies are meat-red or "decay" purple and emit a revolting fetid stench. Some of the aroids are in this group. The huge one called devil's tongue (*Hydrosme rivieri*) from Indo-China is grown as a curiosity in warm climates. Its 8-foot flower stalk holds a greenish-purple spathe within which is a spadix—a cylindrical stem on which is inserted hundreds of tiny flowers. Its odor is distinctly rotten. Moreover, the temperature of the spadix is several degrees warmer than the air, and it oozes a dark fluid. The carrion flies attracted to it crawl over the spadix, often laying eggs upon it, and effect pollination in the process.

The well known *Stapelia* is another bad-smelling flower attractive to bluebottle flies. Its star shape and, in some species, decorative markings make it fascinating in a lurid way. The giant species *Stapelia gigantea* has such a strong odor that a plant in bloom in a greenhouse in Wyoming will bring the flies out of hibernation in January during sub-zero weather. They arrive at the flower almost as soon as its petals unfold. A grower who has observed this phenomenon for several years says the odor of the plant is not just of rotten meat, but of a goat that has been dead for several weeks, so perhaps it is not surprising that it awakens flies from their winter sleep.

An unprepossessing-looking orchid that grows as a vine in Borneo, *Bulbophyllum beccarii,* sets what might be the record for foul smells. It has hundreds of tiny reddish-yellow flowers inserted on a cylindrical stem. The odor they emit has been likened to a herd of dead elephants, and it is recorded that persons studying the plant in an enclosed place fainted from the overpowering stench.

In their search for pollinators, plants haven't missed many possibilities among the ranks of those that creep and fly. Some of the most beautiful flowers are pollinated by bats. When dark comes to the desert, small bats dart around the open flowers of *Cereus* and other night-blooming cacti, hanging by their claws to the flowers and thrusting their noses into the masses of anthers as they lap the nectar with slender tongues. "Bat" flowers have particular characteristics to accommodate the little furry creatures: they have heavy substance and strong stems to support the bats' weight. Their flowers are strongly scented and often broadly cup-shaped so that the bats can fold their wings and curl up inside. The flowers of trees pollinated by bats are held on long stems free from the foliage so that the bats' sonar apparatus can locate them without interference. The calabash, baobab, and sausage trees are three examples.

The sausage tree (*Kigelia pinnata*), a native of Africa, is grown as an ornamental for its fascinating sausage-shaped fruits that hang from stems several feet long. It is cherished in Florida, but bat pollination is uncertain in that area and man must perform the rites. It is a labor of sheer love, and not many are willing to risk life and limb to climb the tall ladders necessary to reach the flowers. In Miami at least one man interrupts his work to visit the trees when their flowers are ready. The arrival of the "sausage-tree man" causes great excitement, and he usually has quite an audience as he climbs around brushing onto the flowers

A bat pollinating a flower of *Agave schottii*, the shin-dagger. (*Bruce J. Hayward*)

the pollen he has brought from another tree. Occasionally he furnishes unexpected entertainment for a garden party.

When pollination is so well assured by straightforward methods, it is not clear why some plants resort to being sneaky or underhanded. But they do. *Cryptocoryne,* for instance, an aquatic relative of the calla lily, entertains beetles in an underwater "night club"—a chamber formed of a waterproof spathe surrounding the spadix that bears the flowers. The plant, which grows entirely underwater, holds the spathe up to the surface to admit its guests. Its "door" opens only in the evening and the morning. The beetles enter in the evening and the door swings shut. They spend the night cavorting among the flowers, becoming thoroughly dusted with pollen, and in the morning the door opens and they stagger out. Then, like habitual nightclubbers, they repeat the performance the next night, leaving the previously collected pollen on the stigmas of another plant.

A most entrancing group of flowers are the aristolochias, among them Dutchman's pipe and the rooster flower. They attract flies and gnats, but their odor is only faintly fetid or fungal. Their petals form a curved tube, the open outer end of which is decorated by tails or flaps or broad ruffles, marvelously striped and mottled. Flies enter the tube and are temporarily trapped, either by downward pointing hairs or by slippery surfaces. To keep the insects active while they are trapped, the flowers have windows to admit light, thin or colorless places along the tube. In their efforts to escape, the flies come in contact with the anthers in the closed rear end of the passage. When there has been time for them to become well covered with pollen, the entrapping hairs shrivel or the slippery surfaces dry up and the flies are released. In the next flower their imprisonment leads to pollination.

There is also the lure of sex. Orchids are notorious in this line. Many of them imitate the female of a certain species of insect so cleverly that the males are fooled into attempting to mate with the orchid—and in so doing they transfer pollen from one to another. Scientists have a word for this, "pseudocopulation." In Australia there is an orchid called *Cryptostylis leptochila* that imitates an ichnumid wasp. The males of this wasp hatch out a month before the females. During that short interval the flowers open and the male wasps spend a happy time "mating" with them, and in going from one to another they transfer the pollen, not knowing that the flowers are not truly female wasps. By the time the real females hatch, the orchids have been pollinated and the male wasps turn their attention to propagating their own kind.

Another "fly" orchid, *Trichoceros muralis,* imitates the female of a different species of fly from that shown in Chapter 1. It also depends on pseudocopulation for pollination. (*Rebecca T. Northen*)

The cattleya orchid has a landing platform nd a well-marked runway leading to the ource of nectar, which lies under the pro-ruding column. Its parts consist of three epals, two broad petals, and a lip formed rom a modified petal. The anther is at he very tip of the column and the stigma eneath it. The visiting bee pushes under he column to get the nectar and comes way with the pellets of pollen on its back. *Rebecca T. Northen*)

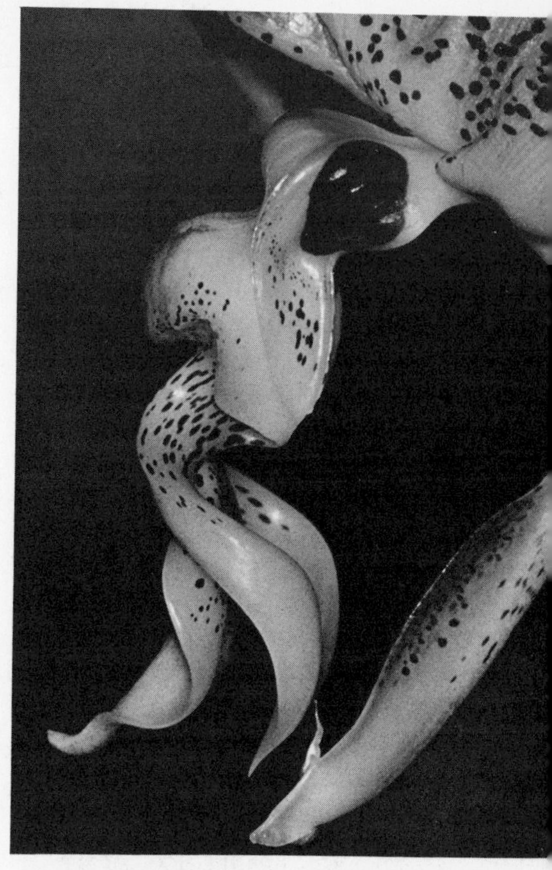

The orchid *Stanhopea* causes the bee to fall down the chute formed by the long column, and the bee is kept from falling off sideways by the horns of the weirdly sculptured lip. As the bee slides through the opening it hits the viscid disc to which the pollinia are attached, and flies away with them stuck to its back. This is *Stanhopea costaricensis*. (*Rebecca T. Northen*)

Cryptostylis might generously be said to fill a real need of the male insects when their own females are not around, but such is not true of other orchids. Members of the European genus *Ophrys* are able to steal the attention of male wasps and solitary bees even when their females are present. Observers have remarked that the females are completely indifferent both to the flowers and to the males' questionable relationship with them. In *Ophrys* the species imitate several different bees and wasps.

In Ecuador and Peru two species of the orchid *Trichoceros* imitate the females of two different flies. In *Cryptostylis* and *Ophrys* the job of mimicking is more obvious to the insects involved than to the human eye, but there is no mistaking the imitation in the case of *Trichoceros*. The hairy body and bristly head, the antennae, and the suggestion of wings are all clear. As the fly makes its mating movements, it comes in contact with the pollinia which stick to it, and it carries them away to the next flower it visits. These two species of *Trichoceros* both stay in bloom for many months, competing successfully with the real flies.

A group of "fast gun" orchids shoots the pollinia at visiting bees, as if to make doubly certain that they won't get away without it. Most notorious are some species of *Catasetum,* which have an additional safeguard to insure cross-pollination by having separate male and female flowers. The male flowers have a pair of triggers, long taillike structures, standing in front of or within the lip. A mere touch on one causes the pollinia to shoot out with great force. A disc covered with fast-setting cement is attached to the stalk that holds the pollinia, and when this hits the body of the bee, it adheres instantly. The bee flies off, and when it visits a female flower it leave the pollinia on the stigma. If you touch a trigger with your finger, you can actually feel the recoil of the flower as the pollinia are shot out; they land with quite a smack.

The ultimate in really wicked treatment is dished out by *Coryanthes,* the bucket orchid. This flower is also the ultimate in fantastic structure. Its lip is molded and contorted into a bucket that hangs down under the column. The bucket is filled with fluid that drips in a steady series of drops from glands at the base of the column. The flowers lure the males of a particular species of bee with a scent so potent that the instant the flowers open shiny green bees come from all directions. The perfume emanates from the base of the lip, and the bees congregate there in mad confusion, brushing the waxy tissues with bristles on their forelegs.

A "fast gun" orchid shoots its pollen at a visiting bee. In *Catasetum macrocarpum* the trigger that fires the gun is the slender taillike structure within the hooded lip. As a bee clings to the upper edge of the lip it touches the trigger and the pollinia are shot upward to stick to its back. (*Rebecca T. Northen*)

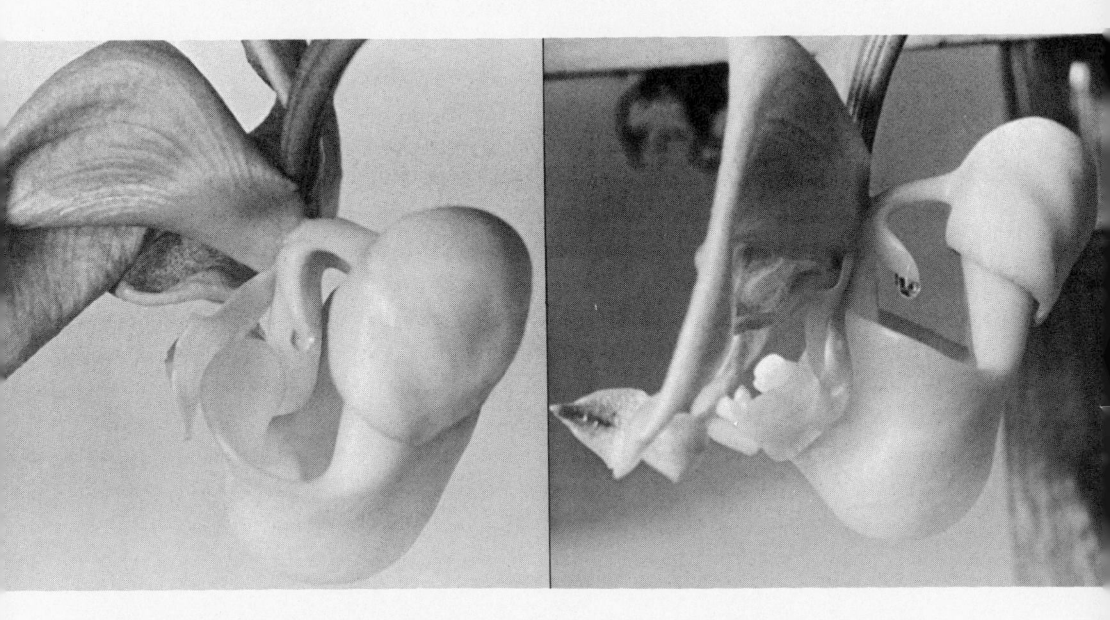

Sneakiest of all is the bucket orchid; the one shown here is *Coryanthes speciosa.* It attracts bees by an irresistible perfume exuded by the globe-shaped upper part of the bucket handle. In a confused condition, the bees lose their footing and fall into the bucket, which is kept filled with fluid dripping from the glands at the base of the column. Unable to fly out, they squeeze through the narrow opening between the tip of the bent-up column and the spout of the bucket, and the first one through removes the pollen. In the view shown here you can look into the bucket and see the little step-stool that aids the half-drowned bee in finding the opening through which it can escape. (*Rebecca T. Northen*)

As they scratch at the tissues, they apparently absorb through pads on their legs an exudate that causes them to behave as if they were drugged or intoxicated. Botanists do not agree as to whether the substance is actually intoxicating, but it does induce chaotic behavior, perhaps affecting them much as catnip affects a cat. It certainly causes them to become unwary, because sooner or later, either by having lost its footing on the waxy surface of the lip or by having come in contact with a drop hanging at the tip of the column glands, a bee is felled into the bucket of fluid. There it flounders, unable to spread its wet wings to fly out. Finally it perceives an exit—a tunnel formed between the spout of the bucket and the tip of the down-hanging column. There is even a stepping stool or a ladder placed below it to aid the bee. The bedraggled insect finds the tunnel quite tight-fitting and it has to exert some effort to pass through. In doing so, it presses against the pollinia and leaves with them glued to its back. The bee flies off, and the flower emits no more perfume that day. The next morning, fragrance again fills the air, the bees again swarm to the flowers, and this time the dunking and exit through the tunnel result in pollination. Ironically, for all their pain and for the valuable service they render, they get no food, only a whiff of an exciting perfume, and a good wetting. Why insects so rudely punished keep returning for more is beyond comprehension, but they apparently make no connection between the alluring fragrance and the ignominious results.

Plants that rely on the wind for pollination do not have the petals, odor, or nectar necessary to those that must lure members of the animal kingdom. Instead, they have relatively small and inconspicuous flowers, but quite large stigmas to trap the pollen, and just one ovule in each ovary. Their pollen is light and dusty rather than sticky. In many the male and female flowers are separate. In the corn plant, for instance, the tassel bears the anthers; each thread of silk is a stigma and style leading to an ovary in which a single kernel will develop.

The wind is an extravagant pollinating agent. Of the pollen whirled away on the breeze, much will settle in far-off places, some will come down on the surfaces of lakes, rivers, highways, and buildings, and much will fall on plants on which it cannot function. In early spring the air is full of the pollen of poplar, birch, oak, ash, sycamore, and hickory. In late spring and early summer, winds bear the pollen of Kentucky bluegrass, timothy, redtop, orchard grass, and Bermuda grass. August and September are months when the air is loaded with the pollen of ragweed, sagebrush, and rabbit brush. Unfortunately, some pollens bring distress to allergic people, and they come to know exactly when the plants to which they are sensitive are in bloom. (The beautiful goldenrod is often savagely rooted out, simply because it comes into conspicuous bloom at the same time as ragweed. But its bright yellow flowers depend on beetles, wasps, and bees, not the wind, for pollination.) Out of this great swirling mixture, the chance of the right pollen reaching the stigma of the right plant is slim indeed, so wind-pollinated plants produce astronomical quantities of pollen to narrow the odds. Just one sorrel plant (*Rumex acetosa*) produces 393,000,000 grains; one of rye, 21,000,000; and one of corn, 18,000,000. In areas of heavy vegetation it has been estimated that in the course of a year 24,300,000 pollen grains fall on every square foot of earth.

Man's ancient history can often be read in the pollen that settled long ago on lakes, ponds, and bogs. Each season wind-borne pollen would drift to the water and sink; each year an infinitesimal layer of sediment would be added to the bottom, gradually filling it with what is called peat. Now scientists can take cores from the peat, wash the pollen from the layers, and read the stories it reveals under the microscope. The pollen of each species differs from that of every other in size and shape and in the delicate and lacy pattern of ridges and protuberances on its surface; each kind can thus be identified. By dating the layers and studying the kinds of pollen they hold, scientists can tell what plants grew in a given area at a particular time.

In very old layers of peat from bogs in Denmark the pollen of oak predominates. In later layers the oak pollen decreases or disappears, and pollen of weeds and cereals appear. The meaning seems clear: only agriculture could bring about such a change. At that point men must have begun to cut down the forests and plant cereals, and weeds thrived in the disturbed ground. Dating of the layers further indicates that the change took place about 3000 B.C., and we conclude that this marked the beginning of agriculture in Denmark, 4,000 years after it had been innovated in Iraq and Iran.

Pollen analysis has enabled botanists to determine past floras and climates of the Sahara Desert. About 12,000 years ago, cedars, oaks, pines, limes, maples, and alders flourished in the area—it was green, lush, and damp. Later the Sahara became drier, and the more drought-resistant olives, cypresses, and junipers shared the region with oaks and pines. Within the last 3,000 years the dry condition has intensified. For a while acacias and grasses populated the area, but they finally disappeared in the waste of desert we know today.

Some aquatic plants have devised special systems to take advantage of, or to function in spite of, their water environment, although some are pollinated just like land plants. The beautiful water lily and the pondweed, which hold their flowers above the surface, are pollinated by beetles and the wind respectively. *Vallisneria spiralis* has a unique system. It grows underwater, producing male and female flowers on separate plants. When the male flowers mature they are set free from the parent to drift up to the surface, where their sepals bend downward to form a tiny boat. Holding erect their two anthers with their large sticky pollen grains, they scud around, driven by the wind and by currents.

Meanwhile, the female flowers are produced singly at the end of a coiled stalk. When they are ready for pollination, the stalk uncoils, pushing them to the surface, where they open and expose three large, leaf-like stigmas. In time a male "pollen boat" will bump into a female "dock" and deposit pollen grains upon it. The stalk of the female then re-coils, pulling it back under water where the seeds will develop. The little male flower, its work completed, dies.

Zostera marina, a grassy plant of the ocean, carries out all of its activities under water. Its pollen grains are long and threadlike, which gives them a better chance to function in the water than if they were round. When they find the stigma of a female flower they coil around it.

Man can substitute for natural pollinating agents. Artificial pollination is the basis for breeding new and superior varieties. Many species that could interbreed in nature are prevented from doing so under natural circumstances. Some are widely separated geographically and pollen from one kind is not likely to reach another. Even though species may grow intermingled with each other they may flower at different seasons or attract different pollinators. Strange pollen, even though it might by itself be able to fertilize the eggs of a certain species, may not be able to compete with pollen from a flower's own kind—perhaps its tubes grow too slowly and by the time they reach the ovules, the eggs are already fertilized.

Man can bridge geographical distances, and he can circumvent different flowering seasons by gathering and storing pollen for later use or by advancing or retarding blooming seasons through manipulation of day length by artificial lighting. He can also select the particular flowers he wishes to cross, and can experiment with many species until he finds those that will accept each other's pollen. Many of our garden and greenhouse plants are man-made hybrids, among them delphinium, roses, irises, tulips, begonias, gladioli, African violets, orchids, and even apples, wheat, and corn. When man has produced a hybrid generation he can choose the superior ones and, by crossing and re-crossing, finally produce varieties far better than their ancestors.

The actual process of making a cross is simple: one plant is selected to contribute the pollen, and another to bear the seed. To prevent any accidental exchange of pollen, the anthers are removed from the female parent before they shed their pollen and its stigma is protected from receiving pollen from any other source. The pollen from the male parent is then placed on its stigma, and the flower is covered with a bag of paper, polyethylene, or gauze. For careful record keeping, each plant is marked. When two species are successfully crossed, their offspring tend to be intermediate between the parents in appearance.

When two hybrids are crossed, or when one is self-pollinated, the mixed ancestry produces a wide variety in the offspring. Some may be quite superior to either parent, others may be inferior. A certain grouping of genes may cause an individual to resemble one of its parents or grandparents, while another grouping may produce a plant quite different from any of them. It is not always possible to know what traits will show up. Much depends on how the genes are distributed to individual seeds. Some genes are dominant for certain traits and make their influence known; others are recessive and cannot show up in the presence of a dominant. For these reasons, a hybridizer must raise a large number of seedlings in order to investigate the whole range of types. Plants that have unusual yet harmonious characteristics—in other words, hybrids that are "new" and at the same time desirable—are sought after, and incidentally can often be patented.

Even though a hybrid will not come true from seed, and self-pollination will seldom reproduce it exactly, it is possible to take progeny from hybrid parents through many successive crosses until a desired set of characteristics does come "true." In this manner superior food crops can be bred, varieties to help lessen the world's hunger. One variety of wheat, for instance, may give a high yield, yet be susceptible to rust disease. Another may not give a good yield, but be immune. The two varieties are crossed and the desired combination of better productivity and resistance to disease sought in the progeny. If it does not show up, members of this first generation are crossed and a second generation obtained. If the desired combination still does not appear, a third generation is made. It sometimes takes many generations (and much time and patience) to produce a particular combination of characteristics and get them to breed true.

The same general method can be followed in producing an ornamental hybrid. The advantage in working with ornamentals is that even if a certain combination of color, size, and shape fails to materialize, the hybridizer may get some other quite unexpected and very beautiful results, perhaps even more attractive than those he had in mind. Disappointments often come his way, too. Not always do two perfectly desirable flowers combine to give worthwhile progeny; sometimes unsuspected poor traits are transmitted along with their good ones. Often, too, some that might be expected to give desirable progeny prove to be incompatible and cannot be crossed.

Occasionally a plant with double the normal number of chromosomes shows up—a tetraploid. If it proves to pass on its good traits, it makes a valuable parent, for it will have double the normal influence for those traits. Tetraploids have also been made artificially by treatment with colchicine, a method employed by many commercial hybridizers. An amateur who wishes to experiment with colchicine would do well to study the methods carefully; it is dangerous to handle, and doses over a very minute amount can also injure or kill the very plants with which he is working.

Since an exact duplicate of a particular hybrid may never occur again in further crosses, a superb variety may be propagated by cuttings (this is standard procedure with carnations and chrysanthemums) or by division. In woody kinds, sometimes more vigorous and hardier plants are obtained when a piece of the selected one is grafted onto a root system of another variety, chosen for its resistance to cold, drought, and root disease. Roses and various ornamental trees are usually handled this way. Nonwoody plants can be grafted too—different species of cacti, for example.

The principle in making a graft is to insert a branch (known as a scion) from one plant into the stem or root of another (called the stock) so that their cambium layers are in contact in at least one place. Cambium is the layer of cells just inside the bark which forms new conducting tissues, that is, new xylem to conduct water, and phloem to conduct food. As the wound heals, the conducting system between scion and stock becomes continuous. A vegetative bud (a bud destined to become stem and leaves) can be grafted in the same manner—a method called budding. Grafting does not change the characteristics of either the stock or the scion. Their cells maintain their integrity and they continue to grow in their own characteristic manner.

The flowering plants have offered their bounty to man freely and generously. He has learned to substitute for their natural pollinators and to manipulate conditions so as to play many variations on Nature's themes. Yet among his rewards are not only the useful products from plants, but pure joy and surprise in the study of the flowers themselves—each offering a story as intriguing as a novel. Even today man is still finding new species with new tales to tell, and no one knows what others await discovery in remote and unexplored regions.

chapter nine

Fruits and Seeds

MUCH of the summer earnings of plants—the carbohydrates, fats, and proteins they make—goes into the formation of seeds. It would be strange indeed if, after such an investment of food and energy, the parents did not provide for their seed's dissemination and safe germination and assure the young plants a reasonably secure start in life. Most do protect their offspring from possible mistakes. Built into the seeds are safeguards—some mechanical, some chemical—that prevent germination under circumstances unfavorable to survival and bring about germination when conditions are right.

A bean is a beautiful illustration of the anatomy of a seed. Soak one to remove its seed coat. Within you will find two fat halves. These parts of the bean, so nourishing for human beings, contain the food stored for use by the seedling. These halves are actually modified leaves, called cotyledons or seed leaves, and plants that have two in each seed are called dicotyledons. (A further distinguishing characteristic of dicots is that the leaves of the plants are net-veined. Walnuts, peas, delphinium, and roses are a few other examples. On the other hand, monocotyledons have but one seed leaf and the plants have parallel veins. Examples of monocots are lilies, irises, orchids, and the grasses.)

If you separate the bean cotyledons, you will find the embryo nestled between them at one end. It is a tiny but fully formed plant whose parts are easy to recognize. There is a short little root, a short stem between the root and the cotyledons (called the hypocotyl), and a plumule—a miniature stem with leaves.

When the bean is planted and provided with water, air, and a suitable temperature, it absorbs water, respiration and digestion proceed, and germination occurs. The root bursts its way through the seed coat and grows down into the soil. The hypocotyl elongates and lifts the cotyledons above the soil. You have probably noticed the little curved stem appear in your garden, pulling the pair of fat seed leaves up after it. As the cotyledons receive light they turn green for an interval, but shrivel as the seedling draws on their stored food. Not all seeds lift their cotyledons above the soil—many kinds remain under ground—but in all cases the growing seedling uses the food they contain, and in all species the plumule becomes the above-ground part of the plant with leaves and stems.

The mystery of what makes the root grow down and the shoot grow up was explained by the discovery that a hormone governs the direction in which they grow. (This is explained in Chapter 11.) Learning that a hormone is the agent should not spoil our wonder at such discerning behavior, for the very control is marvelous enough, and but one of nature's subtle ways of helping plants survive.

A bean seed will germinate equally well in darkness and in light, but after germination those kept in the dark keep the tip of the stem bent over and fail to lift the young leaves into the air. Moreover, the leaves remain undeveloped and folded together. At this stage, if a light is turned on for just a moment or two and then turned off, leaving the plants in the dark again, within a few hours they will have straightened up and spread wide their leaves.

Physiologists have been interested in learning how the various components of white light affect plant processes. They have not accumulated enough data on the blues and greens to give a clear picture of their roles, but they have turned up some fascinating information about the reds.

Our eyes can recognize many shades of red. When white light—that is, visible light—is broken up by a prism, a spectrum of separate colors appears, ranging always in the same order from violet through blue, green, yellow, orange, and red. The wave lengths of the colors are measured in millimicrons (abbreviated mμ—a millimicron is about 1/25 millionth of an inch). Violet, which has the shortest, starts

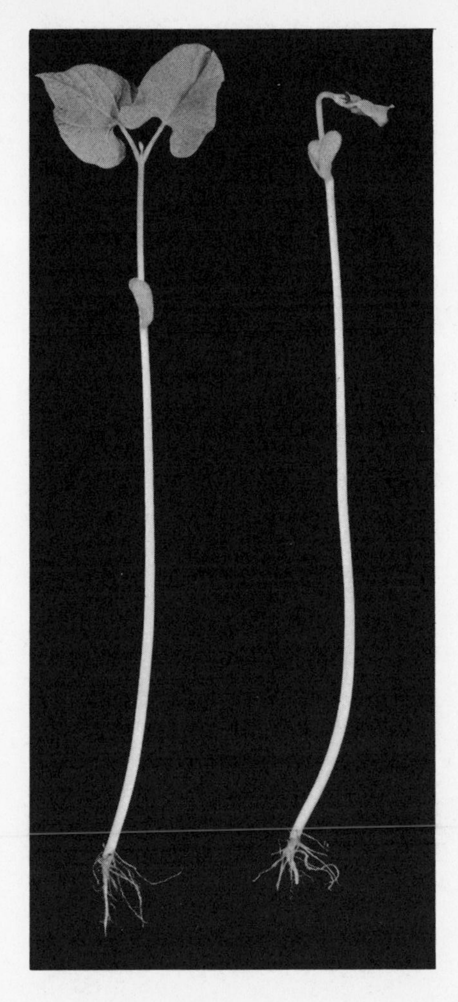

Just two minutes of red light caused the bean seedling on the left to straighten its stem and raise its leaves. The same exposure to far red has no effect on the one on the right. Both were kept in the dark before and after treatment. (*Agricultural Research Service, Plant Industry Station, USDA*)

at 390 mμ. The wave lengths gradually increase as one color shades into the next. The red tones start next to the orange and extend from 650 mμ to 760 mμ. To our eyes the red shades toward orange appear to be lighter and those toward the far end of the spectrum darker. Scientists use the term "red" for those shades with wave lengths from 650 to 695 mμ, and the name "far red" for those from 695 mμ to 760 mμ. Red and far red produce entirely opposite reactions in plants.

A pigment has been discovered, called phytochrome, which controls a plant's responses to the two types of red. The far reds lead to an inactive form; the shorter wave lengths, the reds, lead to an active one. In its active form phytochrome triggers the germination of seeds and controls the development of seedlings. It also plays a role in flowering.

We mentioned that bean seedlings kept in the dark remain bent over and have poorly developed leaves, and that just a moment of white light will snap them out of it. A moment of red light will do the same thing, changing their attitude from one of droopiness to sudden liveliness. But an exposure to far red elicits no response: they remain in their despondent attitude.

Some seeds require light for germination and will not germinate in the dark, among them peppergrass, cactus, and some varieties of lettuce. Their behavior with red and far red light is amazing, and the sensitivity of their seeds even more so. The seed has to be moistened first; dry seed does not respond. If you expose the dampened seed to red light and then plant and water it, it will germinate in the dark. But if you expose it to far red before planting it, it will remain inert. Nor does it germinate if you give it first an exposure to red and then to far red before planting. However, if you reverse the order and give it far red first and then red it does germinate.

The two cancel each other. You can alternate them as many times as you please, and the seed will remember only the last one received, germinating if it was red and failing to do so if it was far red. The shades can be quite close to each other. With the division in the spectrum coming at 695 mμ, the seed will respond to a red of 660 mμ and not to far red of 730 mμ. The seed proves to be able to discriminate between them unfailingly. What a subtle difference there is between a shade that will bring about activity and one that will prevent it!

Nature is no miser, but rather a spendthrift of the most lavish kind. In order to assure the germination of one or two seeds, she produces thousands that come to naught. She has the advantage of not having to reckon cost; the materials she uses are free. But she cannot direct her offspring when they leave home, nor can they direct themselves. They are at the mercy of the elements. Seeds go where winds blow and waters flow, where animals wander and birds fly. Many will come to hostile places where they cannot grow or survive. Only a few will light in spots just suited to their needs. It would seem a risky way to plan for survival, but it works because such tremendous numbers are produced.

In the plant world it is only the children who are free to travel. The adults are too deeply rooted and their work too vital to allow them to move, but they had their excursion in their youth. Now it is their job to produce other wanderers. Even though their travels are not planned, but only wayward, plants have been remarkably successful in finding new homes. Through clever engineering, generous production, and lavish spending, they have spread their colonizers the length and breadth of the earth.

For plants, the problem is to have means for the seeds to be carried away from the parent plants and delivered to places where they can start new life and expand the territory of the species. In order to "market" their products, many plants put them in containers that are delightfully sculptured, richly colored, marvelously scented, and nourishing as well. What man-made package can outshine an apple, a peach, a cherry, or a pumpkin? Man has yet to design one as easy to open as a banana, though he hasn't yet invented a package as hard to open as a coconut.

The container is the fleshy fruit; it is the lure to animals and their reward for spreading the seed. As the "customer" eats the fruit he discards the seeds, inadvertently depositing them in new territory. Of course, some "customers" like to eat the seeds, too, but they often miss a few. Some go through their digestive tract intact, so that enough are scattered to accomplish the plant's initial aim. Plants make no pretense about their products. They present them for exactly what they are. Not only is the package attractive, but it is kept fresh and germ-free for a long time by a usually edible skin and a thin layer of wax. It is the natural wax on an apple that lets it take such a high polish.

In some fruits the ripened ovary wall develops into a hard shell, and the seed within it is the sought-for food. These packages require some labor to open and are therefore likely to be carried home to be worked on at the "customer's" leisure; as a result they are taken some distance from the parent plant. Chipmunks and squirrels perform a great service when they store up seeds and nuts for the winter. Some cached by the former escape being eaten, some buried by the latter are never dug up, and these are thus automatically planted, to germinate the following spring or sometimes years later. Migrating birds transport seeds for several thousand miles, from island to island and even from one continent to another. Viable seeds of *Celtis, Convolvulus, Malva,* and *Rhus* were regurgitated from the digestive tract of a killdeer and a least sandpiper as long as 340 hours after the seeds had been eaten—time enough for flights of several thousand miles.

When fleshy fruits are ready, the "public" is informed by a set of signals. Most fruits remain green in color until the seeds within are mature, and then change to some bold hue that catches the eye. At the same time the flavor changes from sour or bitter to sweet, or to some pleasant flavor that will not pucker the mouth. The few that remain green in color announce their state of ripeness by a softening of substance and by acquiring a desirable scent, both of which can be recognized after trial and error.

Early man and animals learned to select a balanced diet that met all their nutritional needs. If they had not learned this lesson, they would not be on earth today. Many fruits are rich in vitamins as well as minerals and carbohydrates, and a few have a good bit of oil. The food in seeds has a concentration of fats and proteins, which the seedling metabolizes as it grows, and which makes a highly nourishing food for man and animal. In the vast African deserts, where thirst is a danger and the urge to satisfy it compelling, the fruits of many plants are laden with water. During the long season of greatest drought and heat, when few creatures venture out in the daytime, the Bushmen of the Kalahari desert of southwest Africa depend on melons and a few roots as a source of water. They find the fruit amid the sparse, dry grasses, on vines long since withered away.

Not all peoples have available a naturally occurring synthetic sweetener. A West African shrub, *Synsepalum dulciferum,* produces small red berries with the miraculous ability to make sour things taste sweet. It operates in a peculiar way. Unlike our synthetic sweeteners, which are added to a food, this one works on the sense of taste. The local people chew a berry, and for several hours afterward all sour things taste sweet. They use it to make pleasant their sour palm wine and beer and their stale, acidulated maize bread. Without doubt this is one of the most highly prized fruits in that region. The chemical involved has been isolated and has been found to work in the extremely dilute solution of five parts to a hundred million—even when diluted to such an extent, a small amount applied to the tongue is still effective.

Plants mentioned so far use the "soft sell" approach in dispersing their seeds. The "hard sell" used by other species is not so pleasant to contemplate, for here the products are forced on the customer. The containers are not edible and the seeds they contain not especially appealing. The fruits are often pods or podlike and bear spines or hooks or barbed needles designed to attach themselves to passing animals. Later the packages are removed by the animal, and the seeds dropped some distance from the parent plant. They are the burrs or stick-tights that become entangled in the fur of wild animals, in our pets' fur, and in our own clothing. Even though "devil's pitchforks" and the like are annoying to contend with, a magnifying glass focused on them reveals delicate and intricate structures that are fascinating.

Not all plants rely on animal agents to disperse their seed. Some use wind and water, and some have mechanical contrivances to do the job. There are many air travelers among seeds. That of the dandelion is carried away by the slightest breeze by means of its delicate parachute, as are the seeds of arnica and thistle. Milkweed, clematis, mountain mahogany, and fireweed have feathery appendages. The maple seed has helicopter blades that carry it whirling away in the wind. Pine, elm, and ash seeds have wings that turn them into gliders. The cigar-shaped orchid seed is wind-borne because it is as fine as a grain of dust. Tumbleweeds scatter their seeds as the autumn winds roll the dry plants over the land. Seeds or fruits that float, such as those of sedges and the coconut, are dispersed by water.

Oxalis shoots its seed when the pods dry, and a slight touch will stimulate them to do this. *Caragana* has pods that twist open when ripe, forcibly scattering the seed. Pansy (*Viola*), witch hazel, and the wild geranium also toss their seeds some distance. Miniature cannons were invented long ago by dwarf mistletoe and the squirting cucumber. The former pops the seeds to a distance of 50 feet in somewhat the same way you can shoot a slippery watermelon seed by squeezing it between your fingers. The ballistics of dwarf mistletoe seeds have been worked out—they are reported to have an initial velocity of 1,370 centimeters a second and an initial acceleration of 5,000 times gravity! The squirting cucumber sends out a jet of fluid that carries its seeds to some distance.

Dwarf mistletoe squirts its seed with an acceleration of 5,000 times gravity. (*U. S. Forest Service*)

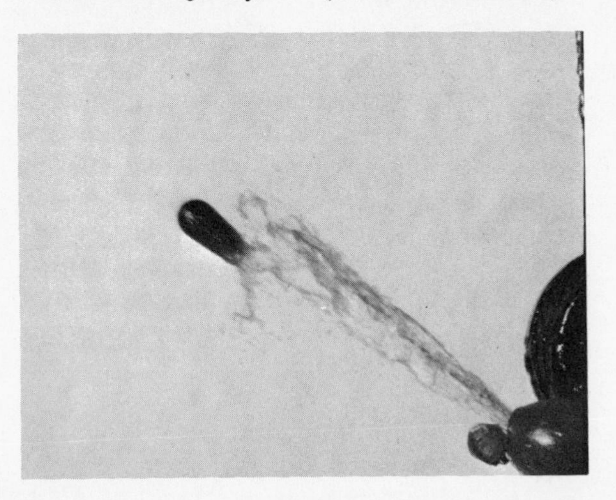

Seeds of most field or food crops germinate promptly after ripening. Left to themselves in our cold climates, they would sprout in the fall and be frozen in the winter. We intervene for them by harvesting the seed and putting it in the ground in the spring.

Wild plants, however, have to shift for themselves. Many have developed delayed-action mechanisms that prevent germination until conditions are right, not only for the emerging embryo, but for the survival of the adult plant. Spruce trees, for instance, scatter their seed in late summer when the soil is still moist and warm, but they do not sprout until the following spring, when they have a better chance to develop.

The embryos of spruce and other plants of similar habit must go into hibernation before they germinate. It is not merely a matter of remaining dormant for the duration of winter, for if the seeds are kept indoors until spring they still will not germinate. The spell they are under can be broken only by a period of cold. Only low temperatures can bring about the internal changes that enable the embryo to recognize spring when it comes and to awaken to its warmth, secure that the time has come to grow. Winter brings about the changes they must have if they are to reach adulthood.

The need for cold before germination is called after-ripening. Many of our cultivated plants require it, among them mountain ash, peach, apple, dogwood, pine, hemlock, rose, iris, and lily of the valley. Their seeds can be after-ripened by being stored in moist peat at temperatures between 40° F. and 50° F. for two or three months. After that they germinate promptly and produce vigorous seedlings.

There are also plants that require cold during their youth rather than as embryos. These are the biennials such as henbane, carrot, beet, and canterbury bell, which do not flower their first season but only after winter has intervened. Cold alters their chemistry, triggering the reproductive cycle so that flowers are produced in the second summer.

Winter wheat is especially responsive to youthful experiences; if not subjected to cold when young, it will not bear grain. Normally the farmer sows winter wheat in the fall and it germinates, sending up shoots that reach a height of a few inches. It then rests during the winter, resumes growth the following spring, and produces grain during the summer. If sown in the spring, it would fail to flower. Special treatment can substitute for winter, however. We can soak the seeds until they barely germinate and then store them in moist sand in a refrigerator for two months. If these chilled and partly germinated seeds are planted in the spring, the resulting plants come into flower and produce grain. Experimental fields in which chilled and nonchilled winter wheat have been planted are most impressive; in the former the plants produce luxuriously, in the latter there is not a single stalk of grain.

A built-in water gauge saves many a desert plant from dying of thirst. Some of our cultivated flowers are not so protected. Rain comes to the desert at uncertain intervals, sometimes only as light showers that dry off in minutes, occasionally as soaking downpours. Seeds that germinate after a light rainfall, as kinds innocent of desert ways might well do, would inevitably be doomed. The native desert annuals, however, remain inactive in such a situation. They require a heavy rain, one that will furnish enough water in the soil to see them through their entire cycle from germination to flowering and seed formation, and only after a real downpour do they respond. Chemical germination inhibitors are embedded in their seed coats. These inhibitors are water soluble, but it takes quite a bit of water to leach them out. A light rain will not do it, nor will several light rains; it takes a heavy soaking to remove them completely. Such a rain incidentally provides enough water in the soil for the plants to grow, and with the inhibitor gone the seeds quickly take advantage of it. Even then, to be sure, they have to go through their life cycle in short order or drought will catch them again. Thus the ones that survive year after year are those that have both the inhibitors and the fortunate ability to reach flowering stage in a matter of weeks. The result is the phenomenon of a desert coming alive with sudden color.

In our Southwest desert the annual species that bloom in summer are not the ones that carpet the ground in the winter. The rains come during a brief period in winter and again in summer, with a long dry spell in between. Why don't all the annuals bloom each time? If they did, some could not survive the greatest heat, whereas others would not do well in cooler weather. They have temperature indicators, sensitive switches that in some species are engaged only by relatively low temperature, in others only by high ones. The combination of low temperature *and* a heavy rain brings the winter species out of dormancy, and that of a high temperature and an equally heavy rain the summer ones.

Even more subtle effects of temperature may prevail. Much scientific research is done with native plants—which often are only weeds—

and clues obtained from them can then be tried on plants of horticultural value. A fascinating bit of information was found in studies of germination of *Senecio vulgaris,* an insignificant relative of the dusty millers of our gardens. When its seeds were germinated at 77° F. its flowering was strikingly more vigorous than when they were germinated at either higher or lower temperatures. Whether other plants may have a similar reaction is not yet known. If some do, it may be possible to obtain extraordinarily luxurious flowering merely by giving the seed beds a temperature different from the normal.

Some seeds depend on light or the lack of it. Bluegrass, lettuce, and many bromeliads and begonias must have light to germinate. They have such tiny seeds that if they germinated at some depth the small shoot would not reach the surface soon enough to start making food for further growth. Other species germinate whether they are put just under the surface of damp soil or on it, although they get a better foothold when they are lightly covered. In contrast, some other seeds need complete darkness, and this is a little hard to explain (not that we can ever expect to "explain" every whim of nature). *Phacelia,* a very lovely wildflower, and many members of the lily family, including onions, will not germinate if light strikes them after they are planted. Perhaps they make some demands of their environment that we have not discerned.

The flowering of some species has long been known to be governed by the length of day. We are just discovering that some seeds are likewise sensitive to daylength. *Begonia evansiana* and birch, for instance, germinate only during long days. It is just the opposite for *Veronica persica,* which requires short days. In this way they may distinguish seasons and germinate only at the beginning of the one favorable to them.

Seeds long preserved in cornerstones or purposely buried in bottles have surprised and pleased later generations by germinating after fifty, seventy-five, or a hundred years. The builders of the Nuremberg City Theater placed seeds of oats, barley, and wheat in its foundation stone in 1832. When it was torn down in 1955, 123 years later, the seeds were planted. The wheat was no longer viable, but 22 percent of the oats and 12 percent of the barley germinated. Seeds of a South American canna were found in a 550-year-old tomb in Argentina. They were inside walnut shells strung on a necklace, evidently to form rattles. Only one grew successfully (and it had somewhat lost its sense of up and down) but when planted, it germinated just as quickly as its modern counterparts. Until just recently, the record for longevity was held by 1,000-year-old (some say 2,000-year-old) seeds of the Indian lotus, which grew after they were dug up out of a bog near Tokyo. There was great excitement when seeds found in the tomb of King Tutankhamen were planted and began to grow, and it was thought that they broke the record of the lotus. But everyone was disappointed when they turned out to be of a kind of cereal not known in King Tut's day, probably having been carried into the tomb by rodents.

Putting to shame all of these are 10,000-year-old seeds of Arctic lupine, cached in a lemming burrow in the Yukon during the Pleistocene Era, and frozen until they were washed out recently by placer mining. About the size of rice grains, some had cracked coats, but those of others were still shiny and intact. Skeletons of the lemmings were found along with them. The animals were apparently sealed off in their burrow by a sudden catastrophe, a landslide or a fall of volcanic ash. The area is one of permafrost, where only the top few feet of soil thaw in the summer and the ground underneath remains frozen year in and year out. The scientists who studied the situation state that covering an area by a landslide or volcanic ash would prevent its thawing, so that the contents of the burrow would have remained frozen had not chance revealed it. Seeds are very resistant to freezing; in fact, they can be preserved longer under freezing conditions than at warmer temperatures.

But freezing for 10,000 years? The scientists planted them, and six germinated within 48 hours. They grew well (they are still growing two years later), and one of them flowered at the age of eleven months. They have proved to be identical to modern Arctic lupine, so it is evident that the species has not changed. But it takes modern ones three years to flower. Did their long freeze in the lemming burrow bring about the early flowering?

A philosophical question arises about these plants. The embryos from which they sprang were still living after 10,000 years, and the plants obtained from them are only a continuation of this life. Are they, then, and not the bristlecone pines, the oldest living things on earth? The bristlecones look their four millenniums; these fresh young plants look like infants, and in fact *are* infants in their present condition. Should credit go to those who have weathered the vicissitudes of life, who have survived its storms and stresses for 4,000 years or more? Or can infants claim the record even though they were preserved as

Seed of the Arctic lupine, frozen for 10,000 years, germinated and developed into these healthy plants. (*Canada Department of Agriculture, Research Branch*)

Four thousand years ago a seed of a bristlecone pine germinated in this hostile environment. (*U. S. Forest Service. Photo by Daniel O. Todd*)

embryos and only now have begun to experience the problems of living? The 10,000-year-old lupines are an enigma. In years, perhaps, they must be considered the oldest, but they definitely belong to the younger generation.

The protection the seed coat affords is one factor that promotes longevity. It keeps the contents sterile and dry. Some coats are so resistant to water that they can be soaked for months, even years, without taking up enough to reach the interior. Coconuts can float with the ocean currents for long periods and reach land before germinating. The germination of several species seems to be promoted by the heat from a fire; a burned-over area becomes rapidly repopulated by plants, some of whose seeds were already in the area but had been prevented from germinating by an especially impervious seed coat. Of course, if they were reached by the fire itself, they would be consumed. But in spots where they receive only the heat, their coats are cracked and shattered. Water can then penetrate and the seeds germinate. In garden varieties speedier germination can be brought about by piercing or nicking the seed coat.

The embryos of some plants are remarkably resistant to heat. Seeds of some species of *Ceanothus* have impermeable coats that can be softened by boiling for five to twenty seconds, with no harm to the embryo. The tolerance of seeds of *Ceanothus cordulatus,* one of a group of ornamental shrubs, is amazing; they remain viable after being boiled for twenty-five minutes.

Orchids, which make such a show of luring insects to pollinate them, provide little for their seeds. As fine as dust, and carried by the wind, the seeds contain practically no stored food. The embryo isn't even completely developed. But the orchids have a partnership with fungi to provide what the parents neglect. Seeds that become lodged in humusy bark crevices, or in sheltered spots among the mosses and lichens along branches, come in contact with fungi growing in the organic matter. The fungi release sugar which the orchid embryos use to complete their development and their first stages of growth. When orchids are grown in cultivation, man has to supply the sugar. The seeds are sown in a flask or bottle on an agar gel containing sugar and the usual plant nutrients. There, during the next six or eight months, they develop into husky little seedlings.

Only a few seed plants are parasitic, among them *Striga,* one that grows on the roots of corn. It depends on corn roots for its survival and will not grow on any other plant; the seeds do not even germinate unless corn roots are present. A chemical, a rare ketopentose sugar released by the roots of corn and not by any other species, triggers their germination. And among all the roots that may be growing in the same soil, the *Striga* seedlings seek out the roots of corn and become anchored to them. Mistletoe is another parasitic seed plant, but it is not so selective as *Striga*.

If plants can be said to have personality, they can be divided into two types: the reckless kind that gambles all its seeds at once, and the more wary, cautious sort that keeps a sensible reserve. Marigold is a spendthrift, sweet clover a miser. If you plant a hundred seeds of each, every marigold will have germinated within a few days. That's all very well, providing everything remains favorable, but if a drought or a freeze occurs, every seedling will be lost. But of the hundred seeds of sweet clover, about 10 seedlings will have made their appearance in a week's time. Twenty will appear after two weeks, 50 after a month, and 75 after a year. Twenty-five seeds will remain ungerminated. If drought or frost should come after the first week, only 10 clover seedlings will die while 90 seeds remain to sprout later on. If something happens to wipe out the 75 that grew the first year, 25 will still be around. Sweet clover's habit of sporadic germination insures survival of the population.

Actually, both marigold and sweet clover had humble origins. Marigold came from Mexico, sweet clover from Europe. In her home country the marigold didn't have to be much concerned with droughts and freezes, and her human admirers look after her here in the United States. But sweet clover knew hardship during her millenniums of evolution and learned her lessons by experiencing adversity. She has had no help from man. She is none too aggressive and is easily pushed out by crowds. She has had to make her home in wastelands, unoccupied lands, or lands inhabited only by a sprinkling of weeds. The bare ground adjacent to our highways and areas burned-over or denuded by floods offer her a temporary home. There she contributes to restoration of the land, but as the years go by the other plants that invade her territory crowd her out. It is not surprising that she has learned to play it safe. But when she moves on, she leaves behind some of her seeds to germinate at such time as the crowds may be gone again and other sweet-clover plants have a chance to grow.

chapter ten

The Engineering and Design of Plants

HURRICANES have not laid low the De Soto Oak at Tampa, Florida, where De Soto made a treaty with the Indians in 1539. No blizzard has torn asunder the Council Oak at Sioux City, Iowa, where Lewis and Clark held council with the Indians in 1804. Still standing is the Tree of the Sorrowful Night near Mexico City, where Hernando Cortez wept after his army was defeated by the Aztecs in 1520. No disaster has felled the Arkewyke Yew at Runnymede, England, under which the English barons compelled King John to sign the Magna Carta in 1215. The venerable redwoods of Muir Woods, across the bay from San Francisco, have withstood earthquakes that crumpled man-made structures.

Plants must resist the elements and they are cleverly designed to do so. A tree trunk must not only be massive enough to support its own weight against gravity and the forces of wind and rain, but also flexible enough to bend and give so as not to break. The more tender plants must be tough enough to spring back after being bent almost double. Strands of fibers whose tensile strength is almost equal to that of steel are strategically located in stems and leaves, along with a system of hollow tubes. Together they form the xylem, which plays a dual role of conducting water and minerals and adding rigidity. The woody plants—that is, trees and shrubs—have one basic design and the herbaceous or nonwoody ones another, but each has a structure that is functional and perfectly efficient.

Whether it is the trunk or the branch of a tree or a shrub, the stem of a woody plant is basically a reinforced column. The wood forms in concentric rings; each year a new shell of water-conducting cells and accompanying fibers is produced on the outer side of the previous year's layer.

The xylem starts near the tips of the roots and proceeds up the trunk and into the branches, twigs, leaves, flowers, and fruits, distributing water and dissolved minerals to every cell on the way. The conducting cells are cylindrical, like pieces of bamboo, and are placed end to end. Their walls are thick and strong, but there are thinner places on their side walls through which the water and minerals can diffuse laterally. They are most active during the first few years of their life. As they become older, they become increasingly clogged with gums and resins, until finally they cease to conduct at all, remaining intact as heartwood.

That tons of water can be drawn through miles of roots and raised to heights of several hundred feet without benefit of pumps is a remarkable phenomenon, and the speed with which it is accomplished is no less so. In the red maple, water rises about 10 feet per hour, and in the American elm about 125 feet per hour. As the leaf cells use water or lose it in transpiration (evaporation), they draw in more from the nearest xylem cells, pulling on the continuous column of water. The water column can be thought of as a rope which is pulled upward as each leaf cell uses water.

Running parallel to but outside the xylem is another set of tubes, the phloem, which carries sugars, vitamins, hormones, and other substances made in the leaves to all parts of the tree. This is the damp and succulent green tissue you see when you peel off the bark, and is therefore called the inner bark. Because it is so nutritious it is sought by animals such as porcupine and deer. Phloem is active for only one or two years; then its delicate cells are crushed as new rings of wood are formed. When a tree is girdled—that is, when the bark is stripped from it in a complete circle—the phloem is interrupted and food can no longer move down to the roots. Water can still move upward through the xylem, but the roots will eventually die of starvation.

The movement of food in the phloem is not clearly understood, but one theory is based on cytoplasmic streaming. If we examine an active phloem cell under the microscope we see that the cytoplasm (the living fluid in the cell) streams around and around within it, looking for all the world like a tiny conveyor belt. At one end the cytoplasm absorbs food coming from the direction of the leaves, and as it circulates, the food diffuses into adjacent cells. Within a large tree there are billions of these rotating streams, not only in the phloem cells but also in the leaf cells. The streaming of cytoplasm led Thomas Huxley to say, "If such be the case [that is, the streaming of cytoplasm] the wonderful noonday silence of a tropical forest is, after all, due only to the dullness of our hearing, and could our ears catch the murmur of these tiny maelstroms, as they whirl in the innumerable myriads of living cells which constitute each tree, we should be stunned, as with the roar of a great city." It should be noted that the little conveyors operate without pulleys or rollers, their motive force that of life itself. Surely they are one of nature's most ingenious contrivances.

The cut end of a log reveals the trunk's structure. The wood and bark are clearly distinguished. Each ring evident in the wood represents one set of xylem, a year's growth, the outermost being that of the year the tree was cut. (The phloem does not survive to show up as rings.) Each annual ring has two bands; the light part, called spring wood, and the darker, summer wood. Xylem cells formed in the spring when growth is most rapid are large and relatively thin-walled, while those made in the summer are smaller and have thicker walls. In the accompanying picture of a microscopic enlargement of annual rings in a section of a

Microscopic section of wood shows the xylem cells, large in the spring wood, small and closely packed in the summer wood. (*U. S. Forest Products Laboratory*)

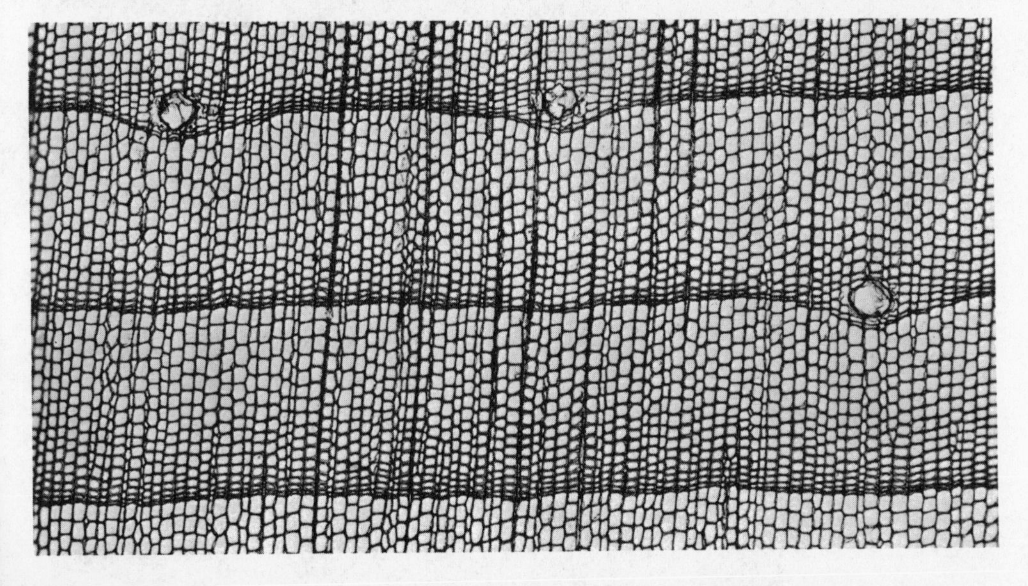

tree, you can see the light and dark portions of each ring, and also that some rings are wider than others, having been laid down during years of optimal growing conditions—abundant rain, plenty of sunshine, and congenial temperatures.

The number of rings indicates the number of years of a tree's life, and it is fortunately possible to count them without cutting down or even injuring the tree. A slender core is simply removed from the trunk with what is called an increment borer. The rings can then be counted, from the outermost to the center, shedding some light on the weather in past years as well. Years of rain or drought are reflected in wide or narrow rings. The pattern of varying rainfall— and hence ring width— through decades and centuries is never quite repeated, so that a tree-ring chronology can be set up and used to date logs used in the construction of ancient buildings.

In any one region, as in our Southwest, each series of rings becomes recognizable as representing certain years, those from 1850–1865, for

To tell the age of a tree a core is removed from its trunk. You can see the rings on the sample being withdrawn. (*U. S. Forest Service*)

example, or 1643–1654. With these as base, ever older series can be put in their place. Cut timbers found in Indian ruins have been dated by this means, so that tree-ring chronology has become an important aid to archaeology. Tree-ring studies in Egypt have helped date ruins in that area. In very old trees, the series of rings are matched with those already known until a starting place is found for the identification of still older rings. In Chapter 14 we'll show how living bristlecone pines have been dated back 4,900 years.

Tree rings also reflect conditions in individual forests and are a useful guide to foresters and timber men. When trees are crowded too closely, they compete for light and water, to the detriment of all. Their rings become narrower. Thinning of the forest is immediately reflected in more rapid growth and wider rings. Trees infested with insect enemies also grow poorly; eliminating the pests gives them a boost in health and the rate of growth increases appreciably.

The bark consists of outer and inner bark. The inner bark, as we have just said, is the juicy phloem; the outer, called cork, is dry and dead. Cork is waterproof and keeps the plant tissues from drying out. It is a shield against insects and disease and may lessen the hazards of forest fires, for it is a good insulator. If the bark is thick and the fire not too hot, the living tissues within the tree may survive.

The cells that make new xylem and phloem lie between the inner bark and the wood—that is, between the phloem and xylem. This is the all-important cambium, which is just one cell wide. During the growing season it forms phloem on its outer side and xylem on its inner side, increasing the diameter of the trunk and the branches. There is also a layer of cells, called the bark cambium, just between the outer bark and the phloem, which forms a new layer of bark each year.

Plants are subject to vascular diseases, many of which are caused by fungi that plug up the conducting tubes and thereby kill the plants. In the chestnut-tree blight a fungus invades and kills the phloem and cambium. In the Dutch elm disease a fungus invades the xylem, and the leaves wilt.

Roots, designed like cables instead of like reinforced columns, are more flexible than the trunk. As a tree's crown sways in the wind, the roots are alternately loose and taut, like the cables of a ship rocking at anchor. The roots increase in length each year, and during any one growing season they collectively roam about as far as the average person travels on his vacation. They progress through the soil, tapping new sources of water and minerals. The finest roots have tiny thin-walled protrusions called root hairs, which reach between the grains of soil and absorb water and minerals. The number on a single tree runs into many millions, and their total surface into thousands of square feet. Root hairs move the water and dissolved minerals into cells within the root, where it enters the xylem tubes and is conducted upward. The roots in turn receive food materials brought from the upper parts of the tree through the continuous tubes of the phloem. Roots of different species may occupy different zones in the soil, for some lie close to the surface and others go very deep. For instance, those of the elm may be confined to the upper 4 feet, whereas those of some oaks may penetrate to depths of 15 feet or more.

Leaves are the food-making factories. They are held horizontally so that their broad surface receives greatest benefit from the sun's rays, the source of power for photosynthesis. They are subject to bending stress, and their design is comparable to that of a well-planned bridge. The veins, composed of xylem and phloem, are the joists that enable them to support their own weight and that of rain or snow. The expanded part of the leaf (the blade) can turn on its stem (the petiole) and so avoid receiving the force of the wind broadside. Strengthening tissues along the leaf margins prevent their being lacerated by winds that would rip a flag to shreds.

The transparent leaf epidermis is coated with wax that retards water loss. It is necessary, however, to have openings in the covering to allow carbon dioxide and oxygen to pass through. Tiny ventilators are spaced in the epidermis, usually on the lower surface but sometimes on the upper as well. These are the pores, or stomata, whose opening and closing is regulated by guard cells. The guard cells open the pores during the day and close them at night. They also close them during periods of drought or on very hot days, thus lessening the escape of water vapor from the internal cells and preventing wilting.

Within the leaf are the food-making cells, usually referred to as the mesophyll (mid-leaf) cells, and they contain chlorophyll in small granules called chloroplasts. In plants whose leaves are deciduous, and those which do not have to go through periods of drought or cold, the epidermis is thin and the mesophyll cells are arranged quite loosely with abundant air spaces in between them. In evergreen species and in desert plants and epiphytes the cells are more compact. Also the epidermis is heavier and the wax coating thicker.

Compare the two cross sections of leaves shown here. The first one is of a water lily, the second of a pine needle. In the water lily, the

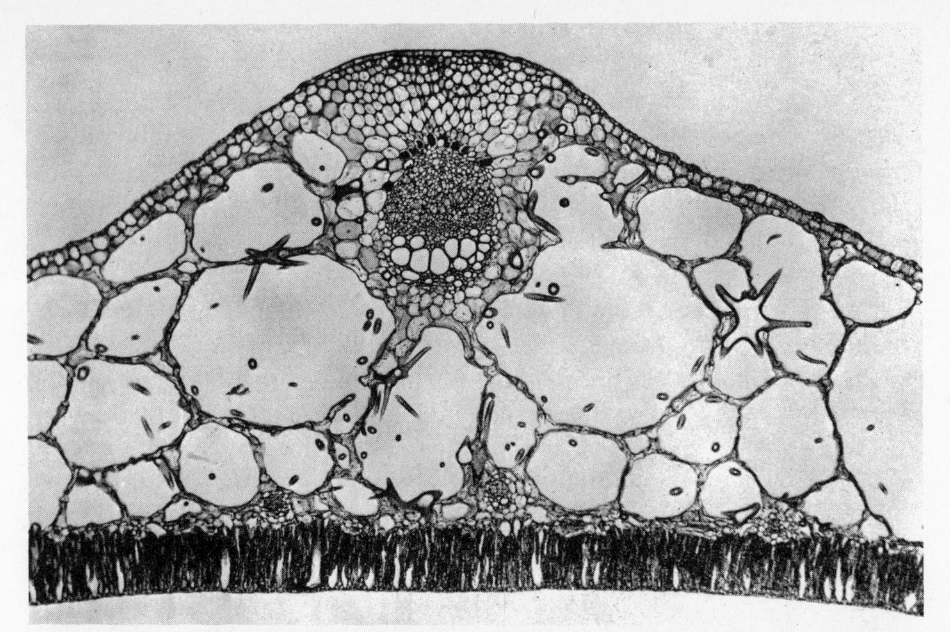

The internal structure varies in the leaves of plants from contrasting environments. The upper one is that of a water lily, which has a thin epidermis and large spaces for the storage and conduction of air. Its food-making cells, containing chloroplasts, are concentrated on the upper surface where they receive light. The thickened area in the lower part is a vein. The much more compact leaf of pine is shown in the lower picture. Its thicker epidermis helps resist drying. The stomata are sunken in pits in the epidermis, and the guard cells can be seen at the bottom of each. The dark irregularly shaped cells contain the chloroplasts and carry on photosynthesis. The central area carries a double vein, and the round empty areas are resin ducts. (*Both photos J. Limbach, Ripon Microslides*)

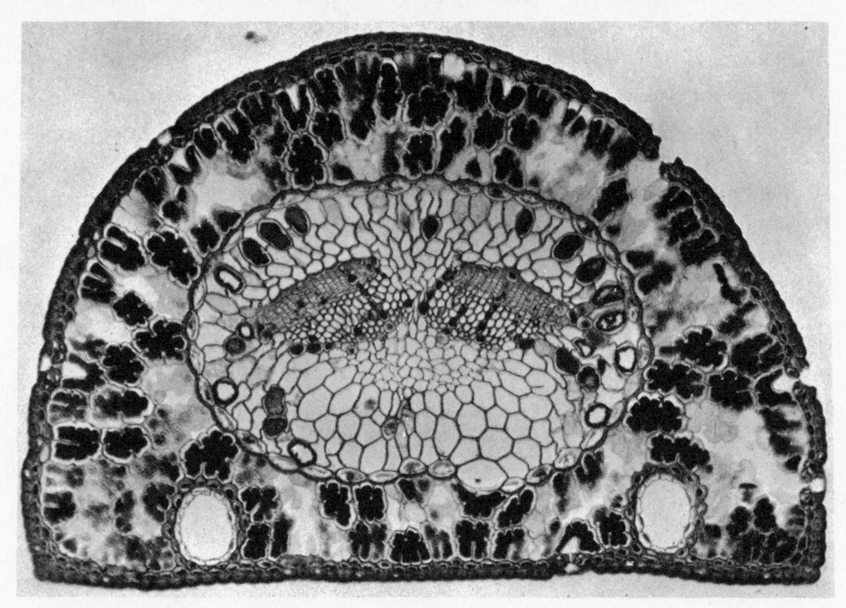

row of tiny round cells along the upper and lower surfaces is the epidermis. The columnar cells along the top under the epidermis are the food-making cells, concentrated where the light can reach them. The large spaces throughout the rest of the leaf are filled with air, (some of which is conducted to the underwater parts of the plant). The region of heavy cells at the bottom is a vein. In the pine needle the epidermal cells are larger. The stomata with their guard cells show up nicely at the bottom of pits or openings in the epidermis. The large, dark, irregularly shaped mesophyll cells are filled with chloroplasts. The oval area in the center is a vein, and the two empty circles at right and left at the bottom are resin ducts.

There is a striking difference in compactness between these two leaves, the water lily that lives in a constantly wet environment and persists only during the summer, and the pine that lives for several years and withstands the cold and drought of winter. In humid regions most plants have broad thin leaves that expose a large area to the light. In arid regions the species tend to have smaller ones, each kind representing a compromise between what will prevent an excessive loss of water and what will allow enough photosynthesis to maintain the plant.

The accompanying X-ray picture of dogwood shows the veins in its large bracts, which, although they are modified leaves, have essentially the same substance as flowers. In some flowers the veins are difficult to see, while in others they are delightfully distinct, often colored differently from, and in contrast to, the ground color.

The herbaceous plants are relatively new additions to the earth's flora, probably having evolved from trees. Two million centuries ago spore-bearing trees—*Calamites, Lepidodendron,* and *Sigillaria*—populated most of the earth's surface. They gave way to the pines, redwoods, firs, spruces, and other conifers that dominated the landscape for more than fifty million years. Then the flowering trees came into being—magnolias (among the most primitive and representing for us today how the earliest flowers looked) and oaks, maples, cherries, and elms—and in many areas they crowded out the evergreens. About one million centuries ago miniaturization began, foretelling the present trend of man's inventions. Just as man's transistors and printed circuits have replaced the miles of wiring in massive electrical equipment, so did the evolution of herbaceous plants replace many of the early giants.

X-ray photograph of dogwood shows the veins in the bracts. Thus are food, water, minerals, and other products carried to all parts of a plant.
(*Oak Ridge National Laboratory News*)

These smaller models now spread over much of the earth's surface. Their great array includes plants that are essential for the nutrition of animals and man. Among the most recent additions to the earth's flora are the sunflowers, mints, orchids, and grasses. The latter are especially important because all cereals are included in the grass family.

In herbaceous plants xylem and phloem are present in lesser amount and in a different arrangement than in trees. Both xylem and phloem are bound together, along with a strand of fibers, into vascular bundles, much like wires in a telephone cable. These bundles are spaced intermittently throughout the stem; the cut end of an herbaceous stalk shows them as little circles of denser tissue, arranged in a symmetrical pattern. They can be stripped out of a stalk of celery, in which they form the "strings." Some species have additional fibers in other places to give added strength, among them the mints and *Coleus,* which have strands at the squared edges of the stems.

In industrial factories the replacement of old machinery with more efficient models results in great productivity. Depreciation and obsolescence are also characteristic of the plant world. The leaves of deciduous trees last only one season; the needles of evergreens are shed periodically, generally after three or four years. Each year, moreover, trees develop additional root tips and root hairs, and new conducting tissue. The functioning xylem is generally less than twenty years old, and the phloem less than two or three years. In annual plants the whole "factory" is depreciated completely in one season. Each year corn, wheat, peas, marigolds, and other annuals build an entirely new structure to produce food for growth and reproduction. By the end of summer, after the seeds are ripe, the plants themselves die, and the seeds develop into new plants the following season.

Stockpiling of supplies is another characteristic of industry that has a parallel in the plant world. During the summer abundant food is stored in the seeds of annuals, in the stems and roots of trees and shrubs, and in bulbs, tubers, and roots of perennials. This food will nourish the plants when growth begins the following spring. Trees and shrubs also take advantage of summer's affluence to make next year's leaves—and some make next year's flowers as well. The leaves that unfold in the spring and the blooms of lilac, cherry, and apple, and the catkins of willow, birch, and cottonwood, all were formed during the previous summer. They were developed just to their earliest stages and held dormant until spring, encased in impermeable coverings to prevent drying by winter's blasts.

Human beings can stretch their physical capacity by exercise and practice. Most important, they can adapt themselves to new jobs, new demands, new stresses. Despite their sedentary existence, plants, too, can become adapted to the rigors of life, and often to unusual demands.

The unsheltered tree, the one growing at timberline or standing alone in a field, is the sturdy one. It has a tough trunk and strong roots, which securely anchor it in the soil, and it weathers every storm. The sheltered tree, one growing in a dense forest, survives only as long as it is part of the community. If the neighboring trees are removed, the first strong wind blows it down. People who build a house on recently cleared land, where only a few trees are left, may well be uneasy when the wind blows, for one of the newly exposed trees may come crashing down through the roof.

A few years ago an Australian scientist compared young, free-swaying trees with some that he had staked so that they could not be moved by the wind. Those that were left free developed stronger, thicker trunks and heavier roots. Those that remained staked for two years were blown down soon after the stakes were removed.

How much movement do plants need to stimulate the growth of stronger parts? Recently an experimenter grew sunflowers on a vibrating machine and found that they became stronger than nonvibrated control plants. Apparently, the vibration made up in some way for natural swaying in the open air.

Plants can even make the extra effort it takes to perform unusual tasks. A pumpkin is normally a tender vine which trails on the ground and allows its fruits to rest on the soil. You could not lift a mature vine and expect it to hold the fruits without breaking. But it is possible to stake a young vine in an upright position so that the developing fruits hang free in the air. As the fruits become heavier with each passing day, the vine becomes stronger to support them, and it can eventually hold those weighing 15 to 20 pounds. Similarly, the leaf stalk of black helebore, the Christmas rose, which can normally support a weight of 12 ounces, will grow strong enough to support 120 ounces—ten times its usual capacity—if the weight is increased gradually.

Children enjoy climbing trees and swinging down to the ground on their branches, and a tree is a fine place to hang a swing. When such activities put an extra burden on the limbs, they actually become stronger. The wood toward the upper surface contracts, so the branch tends to return to its original position, and at the same time, additional wood forms at these points. Naturally, the branch cannot hold a weight out of all proportion to its size and flexibility. A sudden load may break it, as you will often see after a heavy snowstorm.

Tendrils of climbing plants are small working parts that have to support the weight of the whole plant. When a tendril has a chance to function it thrives, but if it remains unemployed it remains small and weak and may even die. The passion vine produces many tendrils, small modifications of the stem. Those that contact a post, a fence, or a branch, will twine around it, increase in size, grow two to twelve times stronger through their lifetime, and support the enlarging vine. Tendrils that do not happen to find anything to cling to simply shrivel away. An extreme example of failure to grow is illustrated by some tropical climbers. Young plants may develop at some distance from anything upon which to climb. If their growing tip does not come in contact with a support, after a certain time the plant loses its leaves, the tip stops growing, and the plant dies.

Certain species such as *Uncaria* and *Strychnos* have hooklike tendrils, and these, like the threadlike ones of the passion vine, grow stronger with use. The employed hooks increase in strength four- to ten-fold and can bear weights of twenty to thirty pounds.

The tendrils of Boston ivy and other species form attaching discs at their tips when they contact a rough surface, but not when they touch a smooth one. *Amphilobium* produces a three-armed tendril with attaching discs at the end of each arm. These discs develop in two or three days when they contact a rough surface, but if they are given a glass post to climb upon the tendrils merely coil around it and discs do not develop. Some species will wrap their tendrils around a glass post but not one covered with gelatin—which, of course, contains a good bit of water and offers less resistance. There is evidently a greater stimulus to the plant from contact with a rough surface than from a smooth one, and little or none with a semisolid one. Not only is a rough surface more irritating to the touch, but it is easier to hold on to. The remarkable thing is that tendrils seem to require the stimulus in order to develop characteristically.

Sweet peas have coiling tendrils, and if you examine those growing in your yard you may find tendrils in all stages of employment and unemployment. The young ones, four or five from a main axis, extend outward as if exploring for a contact. As soon as one finds something to cling to—be it another tendril, a flower stem, or one of the strings supplied for the purpose, it coils tightly around it. Those which do not find a support may remain uncoiled or may finally coil erratically with old age.

The sense of touch in some tendrils is extraordinarily keen. Some very sensitive ones begin to curve five to twenty seconds after a contact stimulus. Those of *Sicyos angulatus,* the bur cucumber (a popular wall cover), are stimulated to curve by contact with a thread weighing only about seven-billionths of an ounce. They are far more sensitive than the human skin, which receives no sensation at all when a thread of this weight is drawn across it.

The tips of some vines and vining flower stems are ready and willing to go to work. In fact, they actually search for it. The tips of the stems perform what is called circumnutation, a rotary movement, which can be loosely likened to "feeling around." As a stem grows longer and the tip continues to feel around, at last it will find the stem of another plant or a post within its grasp. From then on, its rotary movements cause it to coil around and around the object. Such stems coil best in a vertical direction; they seem to lose their sense of direction if presented with a horizontal object or with none at all.

Employment is part of life in the plant world. A plant is really a community in which the individual parts have their functions to carry out, and no member shirks voluntarily. We cannot see most of the work going on, but where we can, we see a conscientious devotion to duty.

Consider the trees in your yard: most of them will live far longer than you, and throughout their long life there is no time when they can let someone else do their tasks for them. Daily they manufacture their own food, obtain their own water, withstand the buffeting of the wind, the heat of summer, and the cold of winter. A tree has to do a full-time job as long as it lives, and—as with ourselves—work prolongs life.

chapter eleven

Hormones

THE sometimes-mysterious behavior of plants is controlled by hormones—the subtle substances that make the roots go down and the shoot go up, that make a plant bend toward the light, and that direct the organism through all major phases of its life cycle. Hormones are extremely potent chemicals. A mere trace can bring about a great response, sometimes even verging on the bizarre. A millionth of an ounce may determine whether a tissue will be tumorous or normal, whether a leaf will age or remain youthful, whether a fruit will fall or cling to the branch, whether a plant will be dwarf or tall. An upset in hormonal balance can result in abnormal growth. A dose somewhat stronger than one that gives the desired effect can be lethal.

It is everyday knowledge that hormones play important roles in the human body. This is equally true for plants: hormones control physiological activities, but like catalysts, they do not take part in chemical reactions. They can also be thought of as chemical messengers, for they are manufactured by certain cells and transported to other regions where they signal or trigger activities. They coordinate the functions of the various parts so that the organism works smoothly as a whole.

The experiments clearly proving that plants produce hormones were remarkably simple. When the tips of the shoots of 5-day-old oat seedlings were cut off, their growth ceased. When the tips were cemented on again with gelatin, the seedlings grew. Cut tips were then placed on gelatin and allowed to remain for a short time, and little blocks of the gelatin were then cut out and placed on the decapitated seedlings. The seedlings responded with a spurt of growth. Obviously, a hormone was produced in the growing tip and moved down through the shoot to stimulate the growth of the entire plant. The hormone was given the name auxin (from the Greek *auxein,* to increase or grow), and is commonly referred to as a growth hormone. The hormone was then isolated and its chemical nature discovered: it was found to be indoleacetic acid.

Plant physiologists began to wonder if closely related compounds might not have the same effect. If indoleacetic acid worked, why not also indolebutyric acid, naphthaleneacetic acid, and so on? Amazingly, they did. Another chemical with a name that indicates it is of a more complicated nature—2, 4 dichlorophenoxyacetic acid—also was found to stimulate growth. (You will recognize it by its abbreviation 2, 4-D.) Other plant hormones are now known, each of which has a specific role.

The auxin naturally made by plants has many effects, all related to its primary function as a growth-promoting hormone. You have probably seen trees growing on steep banks or hillsides, their trunks leaning out at an angle but their tops growing vertically. When the leader of a tree deviates from the vertical—as it might if pushed out of line by snow or wind—the auxin will accumulate on the under side of the stem and stimulate those cells to grow more rapidly, causing the stem to curve upward again.

Something like this goes on in newly germinated seedlings. If the shoot starts off in the wrong direction, auxin accumulates on its under side, and the more rapidly growing cells in that region make it curve upward. Roots are more sensitive to auxin than is stem tissue; auxin beyond a certain concentration causes root cells to grow more slowly rather than more rapidly. Thus an accumulation on their lower sides *inhibits* the growth of those cells and allows the more rapidly growing top cells to make the root curve downward.

You will hear it said that the "shoot seeks the light and the root the dark." But the response can take place in complete darkness, which dispels this idea. You can prove this by germinating seeds in the dark, either in soil or on wet blotting paper, and you will find that the shoots and roots unfailingly turn in their habitual direction. The term geotropism ("geo" meaning earth and "tropism" a movement) is used to describe what appears to be a response to gravity—the root responding positively and the shoot negatively. Actually it is the auxin that is responding to gravity by collecting on the under side of either root or stem.

A plant in a window bends toward the light—a phenomenon called phototropism. When a plant is lighted more strongly from one side than the other, auxin accumulates on the shaded side, where it induces the cells to increase in size more rapidly, causing the stem to curve in the direction of the light. It is not known why the hormone should concentrate on the shaded side, nor why roots are more sensitive than stem cells, but such behavior is certainly advantageous to the plant.

The beautiful symmetry of a spruce tree is the result of hormonal control. The growing tip of the leader produces auxin in large quantities, which moves downward and keep the branches growing horizontally. If the leader is removed, several of the topmost branches will bend upward. If one gets ahead of the others, it becomes the new leader. If two grow at equal rates the tree may become a "double" above that point. A tree with two or more trunks may develop if the growing tip of the original stem is damaged at a young age.

In many plants the hormone produced by the stem tip prevents side branches from forming. Lateral branches do eventually form when the stem grows quite tall and the lower part is freed from the influence of the hormone. A gardener can hasten the branching by removing the stem tip, and this is common practice to obtain bushy, many-flowered specimens of such plants as snapdragon, petunia, and geranium.

Auxin is produced in leaves as well as in the stem tip. As long as the leaves are producing sufficient auxin they remain green and cling to the branch, but when they cease making it they turn color and fall off. (Leaf fall is not restricted to cold regions. In tropical forests deciduous trees replace their leaves a branch or two at a time; they give the appearance of being evergreen, but actually always carry a few bare limbs.)

In temperate regions the shedding of leaves in deciduous species is triggered by the shortening days of autumn. When the length of the days falls below some critical point, the amount of hormone diminishes and in time reaches such a low level that the leaves become inactive. A layer of cells at the base of the petiole (the "stem" of the leaf) forms what is called an abscission layer. Later the cell walls break down, allowing the leaf to separate from the branch.

Other changes take place during this period. Food present in the leaves moves into the stem to be stored for the winter. As the green chlorophyll breaks down, the yellow pigments whose presence was masked during the summer become evident and contribute to the golds of the autumn landscape. Red pigments may be synthesized in waning leaves of maples, oaks, Virginia creepers, and other plants.

Just as a dwindling supply of auxin brings about leaf fall, so also does it cause fruits to drop. As the fruits ripen, their hormone supply diminishes so that the stems become separated from the branch in the same manner as they do with leaves. The fruit is often bruised when it hits the ground, making it unmarketable, so it has become the practice of orchardists to spray the trees with a commercial preparation of growth hormone to help the crop keep its grip on the tree.

Commercial weed killers are not an unmixed blessing, since plants are selectively sensitive to the various growth-regulating chemicals they contain. A concentration that will not harm some species will annihilate others. The chemical 2, 4-D, although it promotes growth in a weak solution, is widely used as a weed killer in higher concentrations. Broad-leafed plants are more sensitive to it than the grasses and cereals. It is therefore used to eradicate dandelions, plantain, and other weeds from lawns and to control weeds in fields. The hormone is absorbed by the plant and brings about abnormal growth, soon followed by death. But as many a home owner has learned, it must not be allowed to drift to other, desirable but still susceptible plants. Rice and other grasses that are resistant to 2, 4-D can be killed with cacodylic acid.

A sad use of 2, 4-D, 2, 4, 5-T, and others—and one that may have far longer lasting destructive effects than originally intended—is its recent use in warfare. In 1967, 200,000 acres of cropland in South Vietnam were sprayed with 2, 4-D to deny food to the Viet Cong. In addition another 200,000 acres in the demilitarized zone were sprayed. Chemical herbicides were also used to defoliate trees over a million acres. By starving the population—women and children as well as men —the American Military hoped to bring about the defeat of the enemy; instead it created an estimated 800,000 refugees. Added to the suffering of human beings and the disruption of their lives is the long-term damage to the land, some of which may suffer serious erosion before it becomes productive again. It is feared that the herbicide 2, 4, 5-T may be the cause of birth defects, now occurring in increasing numbers.

The long-term advantages and disadvantages of control of vegetation should be weighed before chemicals are used. In our own country vast acreages of natural vegetation have been sprayed with 2, 4-D. Where it has been used to kill sagebrush to allow the grasses to grow more luxuriously and provide better forage for cattle, the benefits have proved greater than the harm. The land remains covered, erosion is not a danger, and productivity has been increased. Loss of some wildlife is to be regretted—sage grouse and rabbits among the victims—but the beef produced is needed. On the other hand, the spraying of willows along streams and beaver ponds has not increased the acreage of grazing land sufficiently to warrant the damage. When willows disappear so also do beavers; when beavers are no longer around to keep their dams in repair, the ponds wash out, with consequent disaster to fish and other aquatic life. In eastern forests, trees and shrubs along power lines and many miles along our border with Canada have been sprayed, and now ugly streaks of seared vegetation stretch across the landscape. In the United States 2, 4, 5-T is also causing great concern.

Another type of hormone, kinetin, works together with auxin in cell division and differentiation. The ratio between the two hormones determines whether growth will be normal or abnormal, and whether cells will develop into tissues and organs or proliferate into an unorganized and undifferentiated mass. It was through tissue-culture techniques that kinetin's role was pointed out. Pith cells of tobacco—somewhat unspecialized cells which can become differentiated themselves into tissues and organs if given the proper stimulus—were grown in flasks containing a nutrient solution. When only kinetin was present in the culture flasks, they made no growth at all. When auxin was added, but in greater amount than kinetin, the tissue developed into an unorganized mass resembling a tumor. But when the proportions were reversed and more kinetin than auxin was present, the cells developed into a plant complete with roots, stems, and leaves. When auxin and kinetin are in proper ratio, therefore, growth is normal; when not enough kinetin is present in proportion to auxin, growth is uncontrolled.

Since Ponce de Leon landed in Florida in 1513, the search for the fountain of youth has continued with little success. Scientists interested in geriatrics are seeking formulas for keeping people young, or at least for slowing the aging process. But plants have a potential fountain of youth, at least one that man can make work artificially. After a certain stage in the growth of a plant, nutrients normally move out of the older leaves near its base and into the younger ones near the tip. The older ones then shrivel and die. However, when kinetin is applied to the aging leaves, it reverses the flow: nutrients move out of the younger leaves into the older ones. The older ones become rejuvenated and the young leaves shrivel and die. The old can thus be made to outlive the young.

Is this what would happen if a fountain of youth were discovered for human beings? Perhaps nature is wise. The supply of nutrients is limited. An individual leaf has its share to use and then the share passes on to a newer one. It is only man who can upset their balance by misusing their magic chemical. If man is going to do the same for his own kind, he had better first solve the problem of his limited food supply.

Dwarfs naturally occur in many species in both the plant and animal kingdoms. We have not yet learned how to transform genetically dwarfed animals into individuals of normal size, but we can enlarge smaller than normal members of the plant kingdom by treating them with gibberellic acid, a hormone naturally manufactured by plants. For instance, if dwarf beans are treated with just one billionth of an ounce, they become tall and vining just like pole beans. Apparently the genes of the dwarf variety lack the capacity to form an adequate supply of gibberellic acid, and when the deficiency is compensated for by artificial treatment, the plant grows to equal the normally tall variety. The early growth of seedlings of normal-sized varieties can be increased by gibberellic acid, but it is a moot question whether the extra height improves the plants in general.

Gibberellic acid can also make an annual out of a biennial, causing it to flower the first year when it normally would not do so until the second. The cabbage plants shown in the accompanying illustration were treated wth 1/10,000 gram per week and they flowered the first summer, producing much taller flowering stems than usual. Normally,

Cabbage plants treated with gibberellic acid were induced to flower during the first year. Compare the flowering ones with the untreated controls which formed only heads the first year. (*Michigan State University*)

they would display only the familiar cabbage head their first summer. If you choose to grow cabbage for seed, treatment with gibberellic acid will be to your advantage, but since it wouldn't result in a bigger head (or even a head at all) you will defeat your purpose should you desire cabbages for the table. Of course, you can't eat your cabbage and obtain seed, too.

Every living creature has a time when it is ready to propagate itself, and at that time reproduction is initiated by hormones. In plants, the sudden emergence of flowers is most dramatic. Physiologists can prove that plants manufacture a flowering hormone, which they call florigen, but it has been somewhat of an enigma, for they have not been able to isolate it or determine its chemical nature.

Among the experiments that demonstrate the production of florigen are those carried out with cocklebur. This plant flowers only when the days are short. When grown during long days—of 16 hours, say—the plants become taller and taller, continually forming additional leaves, but they do not come into blossom. However, if just one leaf is provided with short days (by slipping a foil envelope over it after it has received 10 hours of light each day), the plant will bloom. In that one leaf a hormone is produced which moves to the tip of the plant, where it brings about the formation of flowers.

Only the leaves are capable of synthesizing the hormone. If the stem tip is given "short" days while the rest of the plant has "long" days, no flowers will develop.

Grafting experiments also indicate that plants do form a flowering hormone. If a branch from a plant forming flowers is grafted onto one that is not flowering, soon the recipient comes into blossom. The hormone present in the flowering branch moves across the graft union and affects the rest of the plant.

When the hormone florigen has been isolated—and especially if it can be synthesized and produced in quantity—it should be possible to control flowering at will. Manipulation of day length and temperature (also see the next chapter) so far gives the only artificial control, but in plants that can be controlled in this way, it is the altering of day length and temperature that brings about the production of florigen.

Living matter does not give up its secrets easily. Even though scientists have isolated hormones and can give plants experimentally larger or smaller doses and watch what happens, they still do not know what marvelous impetus causes hormones to be produced, or why one part of the plant and not another will respond. While some mechanisms are now understood, the basic mysteries remain profound.

chapter twelve

Calendars and Clocks

AUTUMN is the season of climax, when plants flaunt a glorious burst of color before becoming dormant for winter. Animals scurry in hasty feeding or storing of food to carry them through the months of cold. Birds and animals migrate to winter breeding or nesting grounds. Something stirs in our blood, too, perhaps a call faintly recurring from our own past. The crisp, bright days and blue skies draw us into the country to share the exhibition which nature puts on.

The vast expanse of the countryside no longer wears the green of summer, but brown and gold instead. Where the sun picks up the soft tones on rounded foothills, the texture is almost that of seal skin. Although the grasses and native wild plants are now dormant, below ground rest the plump buds, stems, and roots which will carry these natives through the winter.

On the water of lakes and ponds rest ducks and geese which just a fortnight ago began to wing their way south from Canada or Alaska. Overhead a flight of snow geese may circle, their white feathers glistening like finest satin. The calendars of these migratory birds are as reliable in predicting the coming of autumn as our printed ones.

Where lowlands give way to the foothills, aspen, birch and alder, elderberry and rose heed their own timetables. Even before nights become freezing, they are in autumnal garb and will be left naked before winter comes on. While the leaves are turning to gold or red, important alterations leading to winter hardiness proceed in the buds, branches, trunks, and roots.

Higher up, where deciduous trees are left behind and the forest of spruce and fir takes over, groups of deer are migrating from the high country to their winter homes in the foothills. In mountain streams, brook trout, more brilliantly colored than ever in the summer, make the water boil with their activity. Where in the summer a fisherman might be rewarded with only an occasional catch, the fish are now so numerous that they slither over each other's backs in their race upstream, and occasionally flip out on the bank. They are migrating in search of spawning places—gravelly spots in the stream bed in which to lay their eggs and fertilize them. In the summer they would have darted away from mere shadows; now they are oblivious to any foreign presence.

Plants and animals are seldom fooled by the vagaries of the weather. Just as plants have mechanisms to enable them to wait for and recognize the right time to germinate (Chapter 9), so also do they have a signal to warn them of the coming of winter. In high mountain country, snow may fall in July, but the leaves do not fall, the trout do not spawn, and the deer do not move down. Somehow they know the brief storm will pass and that many weeks of summer will follow. The nights of early autumn are sometimes as warm as those of summer. Yet the migratory birds arrive on schedule and the other events of autumn progress. Clearly, the behavior of plants and animals is not controlled by temperature primarily. They must be prepared well ahead of time to cope with the first blasts of winter. An occasional freeze may catch some unprepared, but in general they are ready when winter sets in.

If temperature is not the stimulus that brings about readiness for winter, what might it be? The signal must be one which is constant from year to year, one that has been and will be the same through millennia. This stimulus, the only condition that does not vary, is day length.

As the days slowly shorten, they reach a critical length that will trip the built-in switches of our companions of field and stream and forest. When these switches have been triggered, events occur rapidly. Food moves out of the green leaves of trees, shrubs, and perennial plants into the over-wintering stems and roots. As we've seen, green chlorophyll disappears and the leaves turn a golden color. Red pigment may form in the waning leaves, which, mixed with the yellow gives shades of red-gold, or even completely masks the yellow. Of greater significance, however, are the internal changes that take place in the branches, stems, and roots, and in the buds that contain next year's leaves and flowers—a lessening of activity and thickening of the sap—which make them resistant to cold and help them retain water when that in the soil becomes frozen. The short days also bring about changes in the physiology of animals, leading to breeding, hibernation, migration, and color changes.

To the ancients, it was a miracle that all that seemed dead should burst into life with the coming of spring. Actually, the new life was there all the time. The winter buds are the promise one year gives to the next—in them lie the hopes of the plants for the future.

But nature does not have much confidence in the whims of the weather. She gives the plants a definite method of differentiating between a January thaw or an unseasonable warm spell in February and true spring. This date on the calendar is marked chemically.

While the over-wintering buds are being formed, a compound called dormin—which inhibits growth—accumulates in them. Cold gradually destroys this chemical. Plants therefore record the passing of winter not by counting days but by means of the diminishing amount of inhibitory chemical. Once it is gone, they await only the warmth and moisture of spring to renew their activity. Species from southern climates require fewer weeks of cold than do northern ones. True spring arrives by February in many areas, and southern species safely resume growth at that time. In northern plants, however, it takes longer for the dormin to be destroyed, so the dormin carries them safely through a false spring. If a lilac is brought into a greenhouse in the fall, it will not open its buds, nor will it open them the following spring when its sisters are blooming out of doors, for its dormin is still present. In its warm greenhouse environment it has not experienced the winter that released the others from dormancy.

To a certain extent we can manipulate the switches that trigger autumn's events. Some time when the leaves are changing color, notice those around street lights; they will still be green when the rest have become yellow or red, and they will cling to their branches when the rest have fallen. Finally, they too will fall because they will eventually be frozen. If we wanted to, we could fool other wild things in late summer with electric lights. Imagine that we could light a lake and its shoreline as a football field is illuminated, and create an artificial

Removal of the bud coverings of willow in December reveals the young leaves and catkins that will open in the spring. (*Henry T. Northen*)

summer day length pattern by turning on the lights before sunset and keeping them on until 10 P.M. The plants would keep their summer garb and the trout would be deceived and would fail to develop their bright colors or start their migration upstream.

Length of day also controls the flowering time of many kinds of plants. Certain kinds bloom only when the days are long, and these are called long-day plants. Other kinds require short days for flowering and are called short-day plants. In addition, there are some that will flower whether the days are short or long, and these are known as day-neutral plants. Among the last are the rose, carnation, African violet, snapdragon, and tomato, which will flower both during the long days of summer and the short days of winter—provided, of course, that in places where winters are cold they are grown indoors.

Chrysanthemums, so highly developed by the Japanese and loved by everyone, are short-day plants. Their marvelous array of colors and shapes and their keeping qualities make them desirable throughout the year for corsages, weddings, and a multitude of decorative uses. They flower naturally out of doors in the fall. But just as the trees around a street lamp are fooled by artificially prolonged days, so also may chrysanthemums be fooled by manipulation of the day length. The flowering of any one species is triggered by a fairly definite day length. Let's take, for example, a kind that flowers with days of 14 hours or less. So precise is its response that if one group is grown with days of 14 hours and another with days of 15 hours, only those with the shorter days will flower. Obviously, they can distinguish that one hour's difference. Without a timepiece could you do as well?

Growers of cut flowers make use of this sensitivity to bring chrysanthemums into bloom at any time of the year. By giving them artificially lengthened days, they hold off their flowering until whatever time they wish them to set buds. At that time they provide them with short days, and the plants promptly proceed to form flowers. Since there are many varieties of chrysanthemums and each has a slightly different day length preference, growers use a day of 8 hours and a night of 16 hours. Since the grower will have learned how long it takes the flowers to develop, he can time their opening quite accurately.

Many other kinds of plants respond to short days. Among them are *Kalanchoë blossfeldiana,* Lady Mac Begonia, *Poinsettia, Stevia,* most asters, and orchid hybrids that have *Cattleya labiata* in their parentage. Some have their internal controls set for days of 12 or 13 hours. *Poinsettia* is one that forms flower buds with days of 12 hours or less. In mid-October the days naturally reach this length, and flowers then begin to form. Two months are required for the development of the flowers, and by mid-December poinsettias are in full bloom, ready to grace our homes for Christmas. (The actual flowers of *Poinsettia* are the hard little nubbins of simple appearance but complex structure that are set in the center of a wreath of gaily colored modified leaves. It is the latter that give the plants their showy beauty.)

Our native plants and plants of our gardens that flower in early spring or later on in autumn are also short-day plants. The choice between spring and fall is determined by other inherent controls that bring about a cooperation between day length, temperature, and the plant's own growth habits.

In the process of working with short-day plants, it was discovered that if white lights were turned on in the greenhouse at midnight when days were short and nights long, the plants failed to form flower buds. Thus it appeared that flower-bud formation depended more on long uninterrupted nights than on short days. In other words, an interrupted night acts like a short night, and the plants respond in the same way as to long days. Now some growers merely interrupt the long nights by turning on lights about midnight, and this prevents flowering just as would giving them long days. Species vary in the length of the interruption that will produce a response. A lighting period of only a minute can do the trick for some. When it is desired to have them start the flowering process, the growers cease interrupting the long night, and the plants then decide they have the short days and long nights they require.

Scientists have found that red and far red light affect the initiation of flowers. You may remember from Chapter 9 that red light includes the shades of red near the orange part of the spectrum and the far red the deeper shades. *Kalanchoë* is a very sensitive plant in many ways and is frequently used for experiments. Being a short-day plant, it flowers normally when the days are short and the nights long and uninterrupted. But when its long night is interrupted with red light for just one minute, *Kalanchoë* does not flower. When the night is broken with far red, it comes into blossom. If exposed first to far red and then to red it does not flower, but if red is given first and then far red, it does blossom. In other words, if you switch reds on the *Kalanchoë* plants, they respond to the last exposure, and red acts as an inhibitor, far red a stimulus, for bloom.

Kalanchoë plants. The plant on the left was grown with long uninterrupted nights and flowered according to its normal habit. That in the center had the nights interrupted by one minute of red light at midnight, which prevented its flowering. The one on the right was given one minute of red followed by one minute of far red. Far red cancelled the inhibitory effect of red and allowed blooming. (*Agricultural Research Service, Plant Industry Station, USDA*)

Long-day plants flower naturally during the summer when days are long and nights are short. Among these are feverfew, *Centaurea, Rudbeckia, Scabiosa, Calceolaria,* the China aster, rose mallow, marigold, spinach, cereals and dill. For these, the days must be longer than 13 or 14 hours. Winter flowering can be brought about in greenhouses by giving them artificially lengthened days or by interrupting the long night near midnight, with white light or red light, but not with far red. For example, *Centaurea* refuses to flower with short days. If you turn on a white or red light in the middle of the night in December, for just a short time, it begins to form flowers, convinced that the nights have become short and the days therefore long.

In addition to calendars, plants have built in clocks, whose mechanism is not yet understood. Some can tell the time within an hour or so, and some almost to the minute, regardless of season. There is a simple marine alga that luminesces (gives off light) at just about midnight every 24 hours. It can be grown in a laboratory where it continues its luminescing habit regularly. Its built-in clock is so accurate that it can tell when midnight comes even when it is deprived of the information of the rising and setting of the sun. When grown in complete darkness, it still luminesces at midnight, night after night, and it doesn't miss by more than a minute or two either way. Such timing is uncanny.

Kalanchoë alternately opens and closes its flowers at 12-hour intervals. If it is kept in the dark, it still keeps to this schedule. Let's say its flowers have just closed when it is put in the dark. Twelve hours later the flowers open; at 24 hours they close; at 36 hours they open; at 48 hours they close again, and at 60 hours they are again open—and this completely without reference to normal day and night.

Not all species have such internal rhythms; in many, variations in light, temperature, and humidity control the opening and closing. Efforts to learn what makes plants open and close their flowers at certain times have yielded some fascinating knowledge about their sensitivity to their environment.

The evening primroses open with the falling temperatures of evening. However, one primrose (*Oenothera biennis*), cannot open at all, no matter how cool the evening, unless the day temperature has reached about 68°. The warmer the day has been, the more rapid is the flower's opening when it cools off.

The gentians are particularly sensitive to temperature. Each kind requires a certain minimum temperature to be induced to open its flowers, after which they will open and close at minute changes, as little as ¼ to 1 degree. Dandelions need a temperature of about 50° to open at all, and about 64° to open to their fullest. The Cape marigold will not open until the temperature has risen to about 62°, and after having opened, it closes when the temperature drops to 48°. This flower was once thought to predict the weather, being open if it was to be fair and closed if rain was coming. However, it is fickle as far as weather prediction is concerned, for it will stay open in the rain if the temperature is warm enough, and close even when the sun is shining if it is cool. In fact, its opening and closing can be manipulated at will, day or night, simply by raising and lowering the temperature.

For a few flowers, humidity is the most important factor. Everlasting (*Helipterum*) closes immediately when the weather becomes damp. Usually, it is open during the drier daytime and closes in the damp of evening. It has been found that it opens when the humidity is 57 percent or below, closes half-way when the humidity goes up to 65 percent, and shuts entirely when it reaches 76 percent.

Among the flowers of field and garden are some that open and close at quite definite times. Some have been given nicknames descriptive of their habits, and others are equally deserving of such distinction. People have for centuries enjoyed working out "flower clocks." If the conditions remain the same for each flower, the clocks are indeed quite accurate, but you can't, of course, blame the flowers if something has happened to put them off schedule.

If you get up early enough, you might see *Tragopogon* (salsify) open at 4 A.M., but by 11 A.M. it will have closed again. For this it is known as Johnny-go-to-bed-at-noon. If you get up a bit later, you can see Iceland poppy open at 5, blue flax at 5:20, and some varieties of morning glory between 5 and 6. These are also early closers; morning glory closes about 11 A.M., blue flax at 2 P.M., and Iceland poppy about 3 P.M. *Portulaca* and marigold are more leisurely, opening between 8 and 9 A.M. and closing between 3 and 4 P.M.

Some water lilies are also a bit leisurely about their opening, the European water lily between 6 and 7 A.M., the Texas water lily between 7 and 8, and the Mexican between 9 and 10. But they don't give us very long to enjoy their beauty, for all of these close about noon. The giant white Amazon water lily is a lover of evening, opening about 5 P.M. and closing at 9. It opens the second day at 5, this time tinged with red, and closes at 9 for good. Visitors to water lily displays are thus often disappointed not to see all of the varieties open at once.

The four o'clocks begin the evening parade by opening at about 4 P.M., followed by the evening primroses and evening star flowers at about 5. From then on many of the night-blooming kinds successively open their flowers, and some close in but a few hours. *Mentzelia nuda,* one of the star flowers, opens about 5 P.M. and closes about 8 P.M. The moon flower, *Calonyction,* opens at 10 P.M. and closes the following forenoon. If you wish to see one of the most breathtaking of all flowers, the night-blooming *Cereus,* you will have to call upon it in the middle of the night, for it opens at 12 midnight and closes at 2:30 A.M.

If these are arranged according to their opening and closing times, as given below, you will see that something is going on at every hour of the day.

Hour	*Name of Plant and condition of flowers*
4 A.M.	*Tragopogon* (salsify) opens
5 A.M.	Iceland poppy opens Some morning glories open
5:20 A.M.	Blue flax opens
6–7 A.M.	*Pachylophus* (morning primrose) opens European water lily opens
7–8 A.M.	Texas water lily opens
8–9 A.M.	*Portulaca* opens Marigold opens
9–10 A.M.	Mexican water lily opens
11 A.M. to 12 noon	*Tragopogon* closes Some morning glories close

12 noon	European water lily closes Texas water lily closes Mexican water lily closes
2 P.M.	Blue flax closes California poppy closes
3–4 P.M.	*Portulaca* closes Marigold closes Iceland poppy closes
4 P.M.	Four o'clocks open *Pachylophus* closes
5 P.M.	Evening primroses open Evening star flowers open *Mentzelia nuda* opens Amazon water lily opens
8 P.M.	*Mentzelia nuda* closes
9 P.M.	Amazon water lily closes
10 P.M.	Moon flower opens
12 midnight	Night blooming *Cereus* opens
2:30 A.M.	Night blooming *Cereus* closes

Vagaries of the weather may cause the floral clock to be off schedule, and this is why it can never be depended upon completely. Also, of course, flowers do not take into consideration such man-made differences as daylight saving time, or the arbitrary division between time zones. The floral clock works best in climates that remain quite constant, in our Southwest, for instance. Here the cactus *Opuntia fulgida* is truly accurate (according to solar time), and you can actually set your watch when it opens at 3 P.M. *Opuntia tesselata* and *Opuntia tetracantha* are just about as reliable, opening at 1:30 and 2:00 P.M., respectively.

It is probably not entirely accidental that most flowers are open when their pollinators are likely to be abroad. Through millenniums of evolution, the inner mechanisms of the plants which cause them to respond to certain conditions of light and temperature have coordinated their movements with those of the pollinators they depend upon.

The evening and night-blooming species are usually pollinated by night-flying moths. There is an interesting exception to this in *Mentzelia nuda,* an evening primrose that opens at 5 P.M. and closes at 8 P.M. Unlike its fellow species, it is pollinated by bees which are just winding up their day's work when it opens. Orchids are especially closely linked with their pollinators, but since most of them stay open for days, weeks, and even months, their opening and closing is not of particular importance. Some of them do have another habit that is rhythmical, a surging and waning of fragrance, some being fragrant only in the early morning, others during the warmest part of the day, and still others in the evening. Since some of them rely on just one species, they must send out their message repeatedly in order to assure pollination.

One last type of activity related to the time of day is called "sleep movements," or drooping or folding of the leaves. Many members of the pea family and some other kinds droop their leaves at 6 or 7 P.M. In *Oxalis* the three leaflets fold in the middle and hang down parallel to the petiole. False acacia (*Robinia*) and false indigo (*Amorpha fruticosa*) droop their leaflets at night. In late afternoon *Mimosa pudica,* especially famous as the "sensitive plant," lowers its leaflets from a horizontal to a vertical position and the leaflets fold together. These are the same movements it makes when it is jarred or stroked.

In many species, sleep movements go on even in constant light or constant darkness. The plants droop or fold their leaves at certain times no matter what is going on around them. All we can say is that it is a built-in habit, carried in their cells from one generation to the next. What frustrates our human brains is that we can see no purpose in it; we cannot associate it with survival or with propagation of the species.

chapter thirteen

Adaptations to Environment

PLANTS have many ways of fitting the niches offered them on earth's surface, of meeting the whims and vagaries as well as the normal conditions of climate, and of overcoming the obstacles of landscape. Sometimes very minute differences in plants' habits make the difference between life and death: the way they make use of light, water, and nutrients, their resistance to some factor, or their ability to make use of something other plants cannot tolerate.

There are forests, grasslands, and deserts all over the world, their location determined by moisture. However, temperature within a region determines what species will grow at what latitude. The kinds of trees in a tropical forest are unlike those of the Great Smokies; species in tropical deserts are different from those of the Great Basin desert. Similarly, grasslands in the tropics are different from those in temperate climates.

Temperatures change both horizontally and vertically—with latitude and elevation—and so does the duration of the growing season. By going from the base of a mountain to its snowy peak you can see the same temperature variation as if you traveled several thousand miles from south to north. In the United States you could follow spring for three months, starting in the south about the first of March and moving north 20 or 25 miles a day. Your way would be lined with trees dressed in fresh young leaves and made fragrant by spring flowers. When you reached the northern border of Maine or Minnesota in mid-May or early June, lilacs and tulips would be just blooming. After spending a couple of months in a cool northern climate, you could follow autumn all the way back down.

From the Grand Canyon's North Rim you can climb on a mule (or walk if you like) and travel in a few hours' time from the climate of Alaska to that of Mexico. From the South Rim, which isn't quite as high in elevation, you begin about at the climate of Central Canada. Before starting off from the North Rim, you have to wait for the 200 inches of snow to melt. In fact, you can't even get there during winter. The South Rim is accessible, partly because there is less snow, but also because the road to the rim is kept open. The illusion of being in a northern clime is enhanced by the snow, of course, but even later in the season the blue spruce of the Kaibab Plateau, above 9,000 feet, and Douglas fir, white fir, and aspen between 9,000 and 8,000 feet, together with wildlife such as blue grouse and ground squirrels, belie the fact that this is not Canada but instead the rim of a canyon in what one normally thinks of as the warm Southwest.

As you dip over the rim of the canyon and start the descent toward the bottom, the vegetation changes between 8,000 and 6,000 feet to what is called a transition zone. Here you pass ponderosa pine, fern bush, and the scrubby Gambel's oak. Jays and chickadees and squirrels soar and frolic around you. Many flowering plants decorate the way. All this is reminiscent of our northern states, Montana, Idaho, and Washington. Proceed lower and you notice the temperature becoming warmer. It is as though you are descending toward southern United States and northern Mexico—sagebrush surrounds you now, and Utah juniper, piñon pine, and blue grama grass. Little rodents such as cottontails and the piñon mouse, and a different assortment of birds live and feed on the plants. At the bottom of the canyon, where the elevation drops to 2,000 feet, you come into the equivalent of the Mexican desert, characterized by the Yucca and other hot-desert life, including rattlesnakes and lizards. Of itself, an elevation of 2,000 feet does not make a desert, but in the Grand Canyon, where water is scarce and where the bare rocks absorb and radiate the heat, desert conditions are created.

In another part of the world, a comparable length of time can take you on a different trip. Let's say you start at the top of a mountain somewhere along the backbone of Central America. Perhaps you visualize tropical regions as uncomfortably hot. This is true enough of coastal areas, but far from the case at high elevations. The Indians of Costa Rica used to fear traveling over the high peaks because of the cold. Their name for one, "Mountain of Death," still clings; legend has it that they used to beat their flesh with nettles to prevent their becoming so numbed that they could no longer move.

The high elevations are usually wet and windy—a combination that makes their barely freezing temperatures more difficult to endure than a less windy, drier, but colder atmosphere. In the chill, biting air at 14,500 feet, only low Alpine-type growth is present, similar to that seen at mountain summits in temperate regions. From the 35° to possible 32° temperatures at 14,500 feet, you experience quite a change as you travel to the 45° or so level at 12,000 feet. Here is timberline with

The altiplano of Peru, elevation between 14,000 and 15,000 feet, is cold and wet. No trees grow here, and the grasses are of quite different species from those of the plains of the United States. (*Henry T. Northen*)

the short, stubby trees quite similar to our wind timber. Moving on down, the temperature becomes appreciably warmer; you can truly feel the change mile by mile. As you approach 9,000 feet and about 55°, you come into the orchid belt. Here the trees are taller, the forests more lush, and epiphytes abound. Going ever lower, the temperature continues to increase until at about 3,000 feet it becomes about 75°. Not so many orchids here—it's too warm for some kinds—but bromeliads love this climate, and ferns, and vines, and aroids. Approaching sea level, the air takes on the quality of a steam bath, almost intolerable to those not accustomed to it, but tolerable to a great variety of plants.

Individual differences in the plants themselves decide which kinds will live at various levels on the walls of the Grand Canyon, on the flanks of a tropical mountain, or on the North American continent from Alaska to Mexico. Where temperatures are not low enough to be a hazard other physiological factors besides hardiness limit what kinds will grow where, but the differences become more crucial when freezing weather is involved. Lack of hardiness prevents many kinds from becoming established in cold regions. The ability to take a few more degrees of cold or a few more weeks of winter gives some plants an advantage over others and allows them to live farther north or at higher elevations.

Native plants survive cold winters by various techniques. Some are annuals whose seeds are resistant to extreme cold. Many perennials have overwintering parts (for example, bulbs, tubers, and rhizomes) below ground. Drought is as much a winter danger as cold—sometimes even more so—and below-ground parts are protected from both. The colder the winter in a region, the higher is the percentage of species with overwintering parts at or below ground level.

In all plants of cold climates, chemical and physical changes occur in the autumn that help them resist winter's extremes. Such changes are known as hardening. Hardened plants can withstand much lower temperatures than actively growing ones. Pine needles may be killed if they are suddenly exposed to 18° temperature during the summer, but during the winter they survive 50° below zero. Whether freezing produces injury depends on whether ice crystals form between or within the plants' cells. Tissues in which water leaves the cells to collect in the interstices suffer less than those lacking this habit, since ice-crystal formation within the cell membrane usually distorts or damages the protoplasm of the cell itself. Species that become winter-hardy often have a thicker, more concentrated sap which does not freeze so readily —you might compare the thicker sap to anti-freeze. Before the cells have become hardened, a cold snap may freeze the cell sap, but temperatures that gradually become colder allow the cells to become hardened before they are subjected to a real freeze.

Many species lack the inherited capacity to become hardened and to resist death as ice crystals form. After the first frost of autumn, beans, marigolds, zinnias, dahlias, salvias, and others wilt, turn black, and die.

Winter drought is a threat because warm sunny days—especially days when the wind is blowing—will extract water that the plant is unable to replace because moisture in the soil is frozen. This is also likely to happen in the spring if the days become warm before the ground thaws. Again, plants that become truly hardened are more resistant to such drying processes.

The intensity of light in a given region depends upon the season, the altitude, the latitude, and weather conditions. The amount of light any one plant receives, though, is related to its position in the community. In a forest, the tallest trees of course get far more illumination than the plants in the understory. Milkweed, evening primrose, aster, fireweed, *Gaillardia,* and other sun-loving species can survive only under the open sky, whereas orchids, pipsissewas, Dutchman's breeches, Jack-in-the-pulpit, wild ginger, trillium, and other shade-loving kinds would yellow and shrivel from the same exposure. There are other species that really prefer full sun but can survive in shade if they have to. The Douglas fir of the Pacific Northwest is a kind that cannot survive in the shade, and we say that it is intolerant. It comes in quickly in cut-over areas where there is nothing to steal the sun from it. On the other hand, the tolerant hemlock can survive and thrive in the shade of the Douglas fir—in fact, the usual situation in nature is tall Douglas fir with an understory of hemlock. Where the forests are managed for timber, the hemlock must be cut over when the Douglas fir is felled, leaving open ground to be reseeded by the sun-loving Douglas fir. Where this is not done, the hemlock will in time replace the Douglas fir to become the climax species.

In foothills and mountains the light intensity is related to the direction in which the slope faces. In the northern hemisphere a south-facing slope receives more light than a north-facing one. With higher light intensity, the south slopes are characteristically warmer and drier than

Many shade-tolerant plants grow in the understory of this cedar forest in Mount Baker National Forest, Washington. (*U. S. Forest Service. Photo by E. Lindsay*)

the north sides, and often support a different plant community. In the foothills of the Rockies, shrubs may populate the south slopes, whereas evergreen forests grow on north slopes. In Ohio and neighboring states a forest of maple and beech develops on north slopes, and an oak-hickory one on south slopes.

To a large extent, the evaporating power of the air in relationship to the amount of precipitation determines a region's plant community. Some water is evaporated from the soil and rock surfaces. A greater amount moves out of leaves and other plant surfaces in a process called transpiration. The amount of water transpired and evaporated will be influenced by wind, temperature, and humidity.

Scientists have made estimates as to the amount of water that would be evaporated and transpired in various localities, assuming that water was available throughout the year. Hence the evapo-transpiration rates are theoretical values. In deserts the ratio of evapo-transpiration to precipitation is two or more. In other words, twice as much water would be evaporated than would fall as rain or snow. In many desert regions, but not all, the soils are fertile, and when irrigated they produce high yields. Since the scanty precipitation does not leach the minerals to great depths, they remain available in the topsoil. In alkali basins without drainage, the salt content becomes too high for crop and ornamental plants and the vegetation is limited to kinds that can thrive in spite of the salt—greasewood, glasswort, salt grass, and a few others. Salinity becomes a problem when soils are irrigated with water containing relatively high concentrations of salts. One of the worst offenders, and the subject of much litigation, is the Colorado River, potential lifeline for the Southwest, but bearer of 6 to 10 million tons of salt a year. The salts accumulate in the soil to which it is applied and after many years the soil may become so saline that crops no longer thrive.

Tall grass prairies develop where evapo-transpiration is equal to precipitation. Here, as in deserts, the minerals so essential for plants remain in the topsoil, which also has a high organic matter content and a good structure. In the United States, Argentina, Russia, and other countries, the rich black prairie soils are ideal for agriculture. The soils are fertile, and precipitation is ample for high yields. But precisely because of their great productivity, only a few remnants of the once great expanses of prairie remain in their original condition; they have all been put into cultivation.

In localities where less water is evaporated and transpired than falls, forests develop. In other words, forests develop where precipitation exceeds evapo-transpiration. Some of the excess water may run off carrying soil with it, but a greater amount moves through the soil to the water table. When such soils are used for cereals or similar crops, the minerals and organic matter are eventually leached out of the top soil, leaving it infertile. As the excess water moves downward in the soil it carries dissolved substances to depths below the reach of roots. When farmed, soils in hardwood forests of the eastern states would, however, remain fertile much longer than soils of a tropical rainforest where the precipitation is four times evapo-transpiration.

All of the factors that have shaped the earth have contributed to the formation of soil. The weathering of the earth's crust by wind, water, and ice, and by heat and cold, has broken up the rocks and washed their particles into the valleys. Erosion and floods have moved quantities of soil from one area to another. Thus the soil in your backyard did not necessarily originate from rock within a few miles of your home; it may be a mixture of materials from hundreds of miles in many directions.

The rock particles that make up soil are of various sizes: the smallest are clay; the next larger, silt; and the largest, sand; and the proportions in which they appear determine the type of soil. Clay soils have a high percentage of the finest particles and a low percentage of silt and sand; sandy soils have little clay and silt; and loam soils have about an even amount of all three. Each type of particle has its own peculiarities and its own role in a soil. Clay, for instance, holds a tremendous amount of water; without some clay in a soil, its water-holding capacity is poor. Sand helps keep a soil open, preventing the compaction suffered by those high in clay.

But soil is more than just rock particles. Organic matter, microorganisms, dissolved minerals, air, and water are other necessary ingredients. In fact, without these, a mere quantity of rock particles cannot even be considered soil, which is a highly complex mixture constantly affecting and being affected by the growth of plants both large and microscopic.

Organic matter is vital to good soil. It increases its water-holding capacity and generally promotes good aeration. It is a rich source of nutrients that are returned to the soil through the action of bacteria and fungi. Black soils contain a large amount of organic matter, and are associated with subhumid climate and tall grass. Today these three factors combine to give us the rich farmlands of our country. Red and gray soils have a sparse amount of organic matter, and in nature these lighter-colored soils go with a cool moist climate and evergreen forests. Similarly, brown soils, a cool and semi-arid climate, and short grass plains are intimately associated.

The acidity or alkalinity of the soil in certain areas limits the kinds of plants that can thrive. Rhododendrons, azalias, citrus, and many others are restricted to somewhat acid soils, whereas some kinds—clover, for instance—thrive better on neutral or slightly alkaline soils.

All plants require thirteen minerals: nitrogen, phosphorus, potassium, sulfur, calcium, iron, magnesium, boron, manganese, copper, zinc, molybdenum, and chlorine. Wherever you see plants growing, whether in fresh or salt water, in soil, on cliffs, or on tree branches, you can know that all thirteen of these minerals are available.

Although plants do not vary in the kinds of minerals they need, they do differ in the amounts they require. Thus certain species thrive only where soils are rich in nitrogen, phosphorus, and potassium, whereas other species may grow where the amounts are sparse. The minerals essential to plant growth were discovered by scientists through growing plants without soil—by a method called hydroponics. Various amounts of minerals were added to containers of distilled water, so that the effects of each on the plants could be studied. Not only did they learn what minerals were essential to normal growth, but what symptoms were produced by excessive amounts or deficiencies.

Where soils are rich in minerals, but not to excess, so also will be the plants, and in such regions animals that graze upon the plants will be amply provided with the minerals they need. In the blue-grass region of Kentucky, soils are rich in calcium and phosphorus, minerals necessary for strong bones. Animals such as race horses raised in this region are sturdy. In contrast to the Kentucky soils, those in many regions of the Atlantic and Gulf Coastal Plains are low in phosphorus, and unless the diet of domestic animals is supplemented, the animals will have weak skeletons.

Oddly enough, the roots of plants don't discriminate between essential and nonessential minerals. They absorb whatever is present in the soil. However, some of those that the plants don't use are essential for animals and man, and it is through the plants that they become available. If iodine and fluorine are present in the soil, they will be taken up by plants and will accumulate in fruits, seeds, and other plant parts. If we eat such parts, our need for iodine (to prevent goiter) and fluorine (to retard tooth decay) may be satisfied. It must be said here that these two elements may also be present in the same regions in drinking water and that man may obtain them from either or both sources.

Sometimes nonessential minerals absorbed by plants may be harmful to grazing animals. Some shale soils of the West have a high content of selenium. Such areas support a characteristic vegetation of *Stanleya,* woody aster, and *Astragalus,* whose shoots contain a high content of selenium. Cattle and sheep grazing such plants may become sick and may die. In the days of the settling of the West, many a rancher went broke on selenium lands, although the cause of the disastrous effects to his livestock was not then understood. Now, selenium soils can pretty well be recognized by the kinds of plants growing on them.

In recent years, an analysis of plants has been helpful in locating deposits of certain minerals—the system could be called "prospecting by means of plants." In certain areas the roots of trees and shrubs may absorb valuable minerals, gold and uranium, for instance. A high content of either in the shoots of plants therefore indicates rich deposits many feet below the surface.

Bogs, marshes, swamps, and ponds, where air is forced out of the soil by the amount of water present, support a different vegetation from well-aerated soils. The kinds that grow in a watery environment are known as hydrophytes and are especially equipped for their particular life. Their leaves, stems, and roots are provided with air passages where air may be stored and conducted from one part of the plant to another. For example, the floating leaves of the water lily absorb air, which is then moved down to the underwater stems and roots. Without such ventilating shafts, or air pipes, the plants would die of suffocation.

Marshes, where the soil is covered by 6 inches to 3 feet of water, are characterized by a grasslike type of vegetation—grasses, bulrushes, cattails, and wild rice among others. Their rhizomes are rooted in the soft ooze of the bottom, while their strong and wandlike shoots extend above the water, subject to the action of the wind. Marshes are the favorite haunts of a great variety of wildlife, including ducks and geese and other waterfowl. Too frequently marshes have been filled or drained, with consequent harm to their many interesting plants and animals.

Swamps support a woody vegetation of trees and shrubs. Deep-water swamps occur on the flood plains of the large rivers of the South, especially the Mississippi. There grow cypress, water oak, swamp black gum, and tupelo gum. They have buttressed trunks and extensive shallow root systems that effectively support them in the soft, water-saturated muck. The roots of cypress, swamp black gum, and tupelo gum develop vertical projections known as knees, sometimes rising 8 to 10 feet above the water. It has been thought that the knees conducted air to the submerged roots, and although recent investigations cast doubt on this theory, it has not been completely discarded. Perched on the tree branches are many epiphytic plants, the most spectacular of which is the Spanish moss, a relative of bromeliads. In the North, swamps are populated by different species: larch, spruce, alder, arborvitae, willow, and red maple.

The line between swamp and marsh is not always clearly drawn. The two intermingle in the great marsh-swamp, the Everglades. In that vast region 40 miles wide and 100 miles long, fresh water from Lake Okeechobee and from various rivers flows under a mat of sawgrass (actually a sedge with its edges sharpened by silica), in some places so dense as to resemble plains. Low islands of rock rise just high enough from the mucky bottom to form hammocks where trees form miniature jungles—wax myrtle, willow, bay, and custard apple. The fresh water, aided by a low barrier on its eastern edge, holds back the salt water of the ocean. Attempts to convert some of the rich black ooze to farmland failed at first because drainage allowed salt water to flow inland. Various methods of control are being tried, but the swamp and its unique inhabitants are threatened.

Bogs are characterized by a floating mat of plant growth, an accumulation of acid peat, and a low supply of nitrogen, phosphorus, and potassium. These factors plus a deficiency of oxygen limit the kinds of plants to those that can tolerate the deficiencies or find a way to gain a supplemental supply. Nevertheless, a rich variety of plants make their home in bogs—orchids, sedges, cranberry, bog rosemary, leatherleaf, Labrador tea, and many carnivorous plants.

The carnivorous plants are particularly fascinating. There are about 500 species, all plants with attractive flowers, able to make their own food as do other green plants. Their leaves are fashioned into traps that capture insects, worms, and other small creatures. They secrete enzymes that digest the animals and the resulting substances are absorbed by the plants to furnish a supply of the scarce minerals. These plants' ability to capture and devour animals has led to weird tales that there are man-eating plants in remote tropical regions. So far such stories must be considered false, because all of the carnivorous plants known are small and the largest creatures they are known to capture are small frogs.

The Venus fly trap, *Dionaea muscipula,* is a small plant that inhabits sandy bogs and pinelands of the Carolinas and neighboring states. Its leaves form an open rosette close to the ground. Each leaf consists of a flat basal portion and a terminal part formed into an active trap. The two lobes of the trap are fringed with pointed teeth and hinged by a midrib. On the flat surface of each lobe are three spinelike triggers; when an insect touches at least *two* of them, the lobes close together in a matter of two or three seconds, imprisoning the visitor. Glands on the lobes secrete juices that digest the insect, taking a week or two to perform the job. After that the trap opens again, to await another victim.

Sundews, *Drosera,* also have active traps. Their leaves are covered with delicate tentacles, often red or red-gold, that secrete drops of sticky mucilage at their tips. Their name means "dewy," and indeed, they glisten in the sun as if covered with dew. The mucilage acts like flypaper. An insect alighting on a leaf becomes entangled, and its escape is further impeded as neighboring tentacles bend toward it, pressing their mucilaginous tips against it. The insect dies and is digested, and after a few days only the indigestible parts remain. Often a dozen or more insects in various stages of capture and digestion can be seen on a single leaf.

There are many forms of *Drosera,* about ninety species in all, scattered all over the world. A number are native to North America, but Australia has the greatest abundance. They grow on damp, mossy ground where it is either wet all of the time, as in bogs, or wet during most of the year. A few have tuberlike roots that withstand a dry season.

In the damp North Carolina woods grows *Drosera rotundifolia,* with a rosette of spoon-shaped leaves, of which the "bowl" is covered with red tentacles. Sometimes they cover the ground almost completely. From Delaware to Massachusetts along the coast in wet sandy places is *D. filiformis,* of entirely different appearance. Its leaves are threadlike, erect, and covered with red-purple secreting hairs. Similar to this is

Carnivorous plants. This tiny *Drosera glanduligera* is only an inch and a quarter tall. Its flowers are bright orange. Even the sepals and stem are covered with secreting hairs. The leaves lie flat to the ground, each a burst of mucilage-tipped rays.
(*Warren P. Stoutamire*)

The beautiful leaves of *Drosera auriculata* are veritable pincushions of ruby-tipped secreting hairs. (*Warren P. Stoutamire*)

D. tracyi, all pale green, which is native to the Gulf Coast from Florida to Louisiana.

The Australian sundews are quite remarkable. One is a plant about 6 inches tall, *D. peltata,* whose slender stem bears from top to bottom little stalked wheels of mucilage-tipped tentacles. Another, *D. binata,* has graceful arms in branched pairs from the top of foot-high stems, gleaming with golden secreting hairs. This one extends from Australia to New Zealand, while the former inhabits India, China, Japan, and the Philippines as well as Australia. It is quite surprising to come upon them growing on sandstone ledges around Sydney, where pockets in the rocks retain water and form miniature bogs. Another lovely species is *D. spathulata,* which forms bright red rosettes on the green mossy background of its woodsy habitat. The giant and pygmy shown on p. 169 are also from Australia. *D. auriculata* grows about 20 inches tall. Its jewellike leaves are green in the center and each silvery secreting hair is tipped with a ruby. The incredibly minute *D. glanduligera* is only an inch and a quarter tall; its "large" orange-colored poppylike flower is but a quarter of an inch in diameter, and even the sepals are covered with secreting hairs.

Pitcher plants rely on pitfalls rather than on active traps. They have modified leaves formed into variously shaped urns or pitchers. They lure their prey by means of nectar secreted by glands scattered on the outside but particularly concentrated within the mouth of the pitcher. Most of them have a very slippery surface within the upper part of the pitcher on which an insect cannot gain a foothold, and many also have slippery downward-pointing hairs. An insect following the trail of nectar enters the mouth and is plunged down the "slide" to the bottom of the pitcher, where it finds itself in a pool of fluid. The hairs prevent its climbing out, and it drowns. Digestive enzymes in the fluid act upon it, and the plant absorbs the resulting substances.

There are several genera of pitcher plants. *Nepenthes* grows in the Orient. Its pitchers are outgrowths of the midrib of the leaf; the midrib extends like a wire hanger and holds the pitcher upright at its tip. They spread their roots through the upper layer of humus on the damp floor of wet forests. Some grow in hot, steamy regions, others up to great heights in fog-bound mountains. There are myriad shapes in a range of rich mottled colors, and each pitcher has a sunshadelike lid. A few creatures have adapted themselves to withstand their carnivorous habits, among them several insects and one spider, which can live and breed in the fluid and can climb out in spite of the barrier of hairs. Among the over sixty species of *Nepenthes,* a few are epiphytic. They ramble through the trees with stems sometimes reaching 60 or 70 feet in length, and dangle their pitchers over the branches.

Darlingtonia is native to north central California and southern Oregon, where it lives on the edges of mountain swamps. It has peculiar little hooded pitchers that look like mustachioed cobras. There is but one species, *californica.* The leaves themselves are formed into tubular pitchers whose tops turn over to form a hood. They are green and white, while the little flaps that form the mustaches are red and green.

Sarracenia inhabits bogs east of the Mississippi River. Its numerous nicknames indicate the associations it held for the early settlers—Indian pitcher plant, side-saddle plant, Devil's boots, forefather's cup, huntsman's cup. The flowers of some are large and attractive and the pitchers are marvelously shaded in tones of green, purple, and red. Most widespread is *Sarracenia purpurea,* which extends from Florida and Alabama north to Labrador and west to Minnesota. It was apparently the first to be discovered, and was described as "the hollow-leaved lavender" in 1672 in Josselyn's *New England Rarities.* He says of it, ". . . the Leaves grow close from the root, in shape like a Tankard, hollow,

Darlingtonia californica, the cobra-like pitcher of California. (*Warren P. Stoutamire*)

tough, and alwayes full of Water . . . the whole Plant comes to its perfection in August, and then it has Leaves, Stalks, and Flowers as red as blood, excepting the flower which hath some yellow admixt. I wonder where the knowledge of this Plant hath slept all this while, i. e. above Forty Years." Another writer says that it was taken to England as early as 1640 and was cultivated by the gardener to King Charles I. This species and *S. rubra* are gracefully urn-shaped and their lids are broad and wavy.

Some other species of *Sarracenia* have pitchers that are tall and straight, more or less trumpet-shaped, with the lids plain or wavy. All are spectacular, and most are well known through the southern states through which they range. Many hybrids have been made between the above two and the species *S. minor, flava, drummondii, sledgei,* and *psittacina,* giving a fantastic array of colors and forms.

A very rare South American genus of pitchers is *Heliamphora,* which grows in wet mossy places on the sandstone mesas along the borders of Brazil, Venezuela, and the Guianas, the region of Hudson's *Green Mansions*. It is not yet well enough known for the species to have been determined completely. One is shown in the accompanying illustration, in which it can be seen, perhaps better than in any others, how the leaves have become modified to form pitchers.

The rare South American pitcher plant *Heliamphora* demonstrates how the leaves are modified to form pitchers. (*Warren P. Stoutamire*)

Utricularia, or bladderwort, is a genus that has both carnivorous and noncarnivorous members. The carnivorous ones are either aquatic or dwell in wet or marshy places, and some aquatic ones have taken up life in trees! They have tiny bladderlike traps along the leaves, branches, or roots. Each bladder is equipped with a valvelike door that opens only inward, and it actually sucks in insects and other tiny creatures looking for nectar or shelter. Within the bladders they are imprisoned and digested. Among those that live in trees is *Utricularia montana,* which sends its stems through the damp mosses on branches in tropical forests. There its bladders, which are filled with water and are only one millimeter in diameter, capture the almost microscopic organisms that live in the humus material. Another, *U. nelumbifolia,* lives in the pools of water held in the vaselike bromeliads. There they capture insects and tiny frogs. They spread by sending out "feelers" or runners that search out other bromeliads, and when one is successful it develops into another plant. The various species have small orchidlike flowers.

It is remarkable enough that plants are able to make do with an environment as they find it, to put up with drought, cold, heat, or excessive moisture, for example. But it is even more remarkable that some kinds have found ways to obtain necessities *not* available in the environment. Thus the development of the carnivorous habit might be considered an ultimate in adaptation. Even more fascinating is the fact that although carnivorous plants come from several different families, they have found ways to attain the same end—that of obtaining needed minerals by digesting animal matter. The achievement is rather mind-staggering. Each line of evolution, that is, each kind of plant, had to accomplish two things—it had to design and construct a successful trap, and at the same time manufacture the appropriate enzymes to work on the captured animals!

chapter fourteen

Forests

WHEN man spread eastward across North America, something over 10,000 years ago, forests were gradually covering a landscape that had been scraped clean by the ice sheet of the last glacial period. Eventually an unbroken forest extended from the Mississippi Valley to the Atlantic Ocean. The early settlers found it undisturbed when they arrived. However, in their need for homes, roads, and farm lands, they began cutting away the trees, progressing ever westward as they cleared the land. Now, only 350 years later, only limited areas are left intact to add beauty to the eastern landscape. Those who can escape from the cities still have the privilege of driving along country roads lined with trees, of climbing wooded hills, of peering into the undergrowth to detect the first sign of spring flowers, of exulting in the green lace of new leaves or rich autumn colors against the sky.

In the North and extending into Canada, vast acreages were covered with pine, spruce, fir, hemlock, cedar, tamarack, birch, beech, and maple, the deciduous species occupying the better soils, and the conifers the poorer, rockier ones. Spreading southward from these lay the central hardwood forest, populated by oak, hickory, ash, black cherry, walnut, and chestnut. In the South the sandy soils supported a variety of pines—longleaf, shortleaf, slash, and loblolly. The moist bottom lands were inhabited by sweetgum, several kinds of oak, elm, red maple, and eastern cottonwood, while bald cypress grew in the swamps. On somewhat drier sites were magnolias and live oak festooned with Spanish moss. These great eastern forests thrived under ideal moisture conditions, with rainfall in excess of 30 inches and a humid atmosphere to support them during their growing season and to protect them from drying out during the season of dormancy.

The Great Smoky Mountains of North Carolina and Tennessee, a wonderland for us today, offer forests that show how the whole East once looked. Not only are the Great Smokies treasured for sheer beauty but as storehouses of geological and plant history, now preserved as the Great Smokies National Park. About 150 kinds of trees—some of them in the most extensive stands to be found in this country—1,400 flowering plants, and hundreds of kinds of mosses and fungi live there. The mysterious haze that rises over the mountains and gives them their "smoky" look is composed of a mist of oil droplets evaporated from the forests' leaves; this is what gives a forest its fragrance.

These forests are older than those to the north which have come in since the glacial age. Moreover, they are home not only to the southern species that have long been native there but to many northern ones that migrated south in front of the glaciers during the ice ages and were left behind when the ice sheet finally retreated. Clothing the many peaks that reach 6,000 feet and over are majestic red spruce and Fraser fir, reminiscent of forests of southern Canada and northern New England. Between 6,000 and 4,500 feet these give way to maple and yellow birch, a kind of forest found in Vermont. Below 4,500 feet in sheltered locations are the cove forests, where fog rising in the valleys and clouds nestling on the slopes enable a lush, moist forest to develop. Mosses carpet the ground and grow on stumps and logs; droplets of water freshen the flowers of the undercover. Forty-five kinds of trees are crowded into these coves alone, among them oak, hickory, yellow poplar, white ash, hemlock, northern red oak, beech, and basswood—familiar species to visitors from many parts of the East and South. In season, the landscape is made gay with the large and colorful flowers of magnolias, rhododendrons, and tulip trees, the showy bracts of dogwood, and purple clusters of redbud. The size some species attain here is greater than anywhere else; for example, tulip trees reach 200 feet in height, and the silver bell, or snowdrop tree grows to be 20 feet around. The undercover reveals, among the many kinds of wild flowers, delightful wild orchids.

Separated from the eastern forests by grasslands are the forests of the Rocky Mountains, and further west, beyond the deserts, those of the Pacific Coast.

Up to timberline, the Rocky Mountains are heavily clothed with trees which add beauty to their eight national parks and fifty national forests. As you ascend a mountain in this region, you encounter Douglas fir and ponderosa pine intermingled with aspen at elevations of about 7,000 feet. The contrast between the dark evergreens and the dainty flickering leaves of the almost white-trunked aspen becomes especially charming when the latter's foliage turns to gold in the fall. Somewhat higher up, pure stands of lodgepole pine appear where areas have been logged over or suffered fires, their trunks straight and branchless almost to their tops. At 9,000 or 10,000 feet, depending on the latitude, spire-like spruce and its constant companion alpine fir clothe the slopes and continue to timberline, where they become dwarfed and matted. Above these rise great peaks of bare rock, the highest of them trimmed with permanent snowfields.

The oldest living things on earth, the bristlecone pines, patriarchs of the plant world, inhabit the southern Rockies, from the Nevada-California border through Nevada, Utah, and southern Colorado, and into northern Arizona and New Mexico. Here they have lived for untold ages at elevations up to 11,000 feet, struggling with wind, cold, and drought, growing sometimes on almost bare limestone slopes with as little as 12 to 13 inches of rain a year. Until the 1950's it had been thought that the oldest trees were the "Giant Sequoias" or "Big Trees," but through the science of dendrochronology, or tree-ring dating, it was found that the bristlecones (*Pinus aristata*) were more ancient. One specimen in the Snake Range of eastern Nevada is known to have started life in the year 2932 B.C., a thousand years before ancient man made the transition from the Stone Age to the Bronze Age in the Old World. It shows the scars of its 4,900 years, and is worn by time. Another, in the White Mountains of eastern California, has been dated at 4,600 years of age, and several more in that area at well over 4,000 years.

No great stands of the bristlecone pines occur; they are found in restricted groves or as individuals, all at high elevations. It is hard for a young tree to get a start. The man who discovered their real age, the late Edmund Schulman, said that it was easier to find very old ones than very young ones. One about 3 feet tall and 4 inches in diameter turned out to be 700 years old. They grow slowly, adding barely an inch to their girth in a hundred years, and they hold their needles for 20 to 30 years.

Their wood is extremely solid and full of pitch. A photographer working among them found his camera gummed up by the amount of resin in the air. The resinous wood resists drought and also decay, so that a tree with rot is seldom found. As with all timberline trees, the top makes fair growth during a series of good years, only to be killed back by dehydration and starvation during bad years. Thus while they make huge growth at their bases (the one called Patriarch is 37 feet around its trunk), their tops consist of many dead spikes. Their trunks and branches become sand-blasted by the wind with great sections of bark and wood actually scoured away. Bark may remain on only a small area, supporting a few struggling branches. Their roots are often eroded out of their rocky foothold. When part of a tree is killed, growth sometimes comes from the base, so that some have developed many successive trunks from a single root system.

So durable are these bristlecones that even after death they resist being reduced to dust. Eroded snags of some that died at about the time of Christ's birth still stand upright, and their artfully sculptured remains add beauty to the landscape. Study of the rings of the dead trees and their remnants have revealed the eras during which they lived.

At present, a tree-ring chronology for the bristlecone pines, through study of both living and dead trees, has been established for 7,100 years. This chronology can even be used to date mere remnants picked up off the ground. Sometimes these remnants show a matching section at the very end of the known chronology, and can therefore push the dates back a few more hundred years. Visitors to the bristlecone forests who pick up pieces for their beauty may be removing the very ones that could add more data to the known chronology, thus depriving us of more knowledge about these venerable trees—as well as the chance to know what the weather was like back in, say, the year 5550 B.C.

The bristlecone pines are preserved in Inyo National Forest in the White Mountains of California. As one looks across the austere slopes, the wraiths of old bristlecones stand out against the not-too-distant Sierra Nevada, whose snowy peaks trap the moisture coming from the Pacific and deny it to the bristlecones. With just a bit more moisture, who knows to what great heights these trees would rise? It's strange, though, that the oldest living things should survive in such a harsh environment.

Ghost of an ancient bristlecone pine flings its arms into the wind high up in the White Mountains of Inyo National Forest. (*U. S. Forest Service. Photo by Daniel O. Todd*)

The most lofty trees of the United States, as well as those most economically valuable, grow in the forests of the far west, rivaling each other in size and grandeur. The West Coast climate is ideal for forests. Breezes from the Pacific temper the climate and bring ample rainfall—as much as 12 feet annually in certain areas such as the Olympic Peninsula. One-third of the nation's saw timber comes from the Pacific Coast, extending from southern Alaska to central California. Ten national forests and six national parks are located in this extensive region, and additional ones are needed to preserve these forests, some of them unique in the world.

From coastal Alaska to Washington, the king of the forest is the Sitka spruce, which lives with, and towers above, western hemlock, reaching its greatest height at its most southern limit in Olympic National Park. Sitka spruce has a huge buttressed base and a shaft as straight as a mast.

The dominant tree of our Pacific Northwest is the Douglas fir, which, though it grows in other areas as well, covers vast regions of Washington and Oregon with dense stands 500 to 1,000 years old. Individual trees set no record in reaching 50 feet in circumference and 200 feet in height. This tree is carefully nurtured as a source of timber; logging practices are designed to allow new stands to start for the benefit of future generations.

On the Olympic Peninsula is found the paradox of a rainforest in a temperate region. With 12 feet of precipitation a year, Olympic National Park is constantly wet. Water drips from the leaves and glistens on the mosses that carpet the ground, clothe the trunks, and hang like beards or curtains from the branches. When a tree falls, it is quickly covered by mosses and ferns that form a green garden and a seedbed for new trees. Tree seedlings that start life on a fallen log send their roots down its sides and into the soil. When the log decays they are left standing on stilts, sometimes 10 feet above the soil surface in a line as straight as if laid out by a ruler.

In this rich environment four of the Northwest's most important trees reach their greatest size. Western hemlock attains a diameter of 9 feet; western red cedar, 21. Sitka spruce, whose diameter here reaches 15 feet, and Douglas fir, which attains 17 feet, both grow to be 300 feet high. The sunlight filtering down through the trees and glinting on the mosses gives an eerie green color to the atmosphere, giving the visitor the impression that he is underwater.

Farther down the coast, from southern Oregon to just north of San Francisco, grow the redwoods. Muir Woods is perhaps the best known area, but the trees reach their climax in a 100-mile stretch south of Eureka, California. They grow inland only as far as the Pacific fog flows landward—a belt of about thirty miles. They have protected their environment from the ravages of time and the elements for a million years. No erosion has taken place where their great roots hold the soil. Half-burned trunks testify to their survival of fire; for in spite of having half their trunks burned away, the trees still live. Old stumps prove to be larger than any surrounding trees, evidence that they had reached a greater age than any that have grown since. *Sequoia sempervirens* is their name, the ever-living sequoias. The name *sempervirens* is apt in another way as well, for new shoots come from old stumps and from burls formed at the bases of old trees. These burls are interesting in that they seem to be a mass of tissue covered with buds, any one of which is a potential tree.

Individual trees reach an age of well over 2,000 years. One that was cut down was found to have been 2,200 years old. Even if, at this great age, they are not the oldest living things, they do reach a greater height than any other trees in the world. Until recently the record was held by the Founder's tree at 354 feet; then the Libbey tree exceeded it at 367.8 feet, and of course both are still growing. What a travesty against nature it is that some people think their wood is of greater value than the living tree!

As you stroll amid the massive trunks you strain to see the sky through the lacy branches two to three hundred feet overhead. Little sun comes through; you cast no shadow as you walk. The soft ground absorbs all sound of your step; the dampness allows no twig to crackle. Only their own seedlings, ferns, and occasional flowering plants can grow in the dense shade.

Greater in sheer mass and older in years are the so-called "Giant Sequoias" or "Big Trees," which live at high elevations in the Sierra Nevada of east central California. The confusion in names, and the fact that this tree is quite different in some ways from the redwoods, has led to the coining of a new name for it, *Sequoiadendron giganticum*. It is shorter than the redwood, with a far more massive trunk. Its branches are sturdy instead of delicate, and its needles are scalelike, resembling those of juniper, rather than flat like those of redwood. The

Sequoiadendron, the Big Tree, largest form that has ever lived on earth. This is the Frank Boole tree in Sequoia National Forest, 112 feet in circumference and 269 feet tall. (*U. S. Forest Service. Photo by Norman L. Norris*)

"Big Trees" do not grow in pure stands like the redwoods, but as individuals among other kinds, or a few together in groves. Amid the splendor of the scenery of the Sierra Nevadas, they tower over their pygmy companions and battle winter storms. Some are 3,500 years old, and one—the General Sherman, which may come close to 4,000 years old—is 30 feet in diameter. The tallest is the McKinley tree, not quite 300 feet. Several other famous ones approach this size. They are protected in the Giant Forest in Sequoia National Park and King's Canyon National Park, and in the Mariposa Grove in Yosemite.

The "Giant Sequoia," or *Sequoiadendron,* and the redwood once ranged over most of the northern part of the world. The few that are left to us now are all that remain since the great ice sheet destroyed their native range.

Until recently, a third member of the sequoia group was thought to be extinct. This is the dawn redwood, or Metasequoia (*Metasequoia glyptostroboides*). Living trees were found in the interior of China in 1947. Fortunately seeds were brought out, and although they have not been propagated extensively, they are extremely fast-growing. Some, planted only a few years, have already reached 40 feet. They closely resemble seedlings of *sempervirens* in their beautiful shape and in the delicacy of their branches and needles, although their wood is softer.

In addition to hundreds of lovely and valuable trees, Americans are blessed with three of the world's most remarkable species. Our opportunity to care for and preserve them is precious. How poor future generations would be without these formidable conifers to teach them about eternity!

In all the world, the wet tropical forests are the richest of all communities in terms of number of species. In these rainforests, up to four times as much water falls as can be evaporated. Where the weather is hot and evaporation is rapid, this requires between 160 and 400 inches per year. At higher elevations, where evaporation is less rapid, 80 to 160 inches will maintain the same ratio. Then there are the moist forests or cloud forests, mostly found at still higher elevations, which receive less actual precipitation—from 50 to 80 inches a year—but which are kept damp by clouds drifting through the trees, rising from the moist valleys, bathing the vegetation, and keeping the air constantly humid.

A cloud forest is a beautiful sight. You can best appreciate it as you look out over a mountainside and watch the mists rising through the trees like wisps of smoke. In all these forests there is no dearth of water, and the temperatures are mild the year around. Vegetation runs rampant. The trees are of so many kinds that not all have been identified. They grow in sometimes as many as four levels or stories. Short species are exceeded in height by the next taller kinds, and those by still taller ones, until finally the tallest, reaching up 200 feet from great buttressed bases, elevate their heads above all the rest. The tallest reach into full sun, while others live in their shade. The shade becomes more dense for each lower story.

In these forests practically everything serves as a perch for something else. Plants grow upon plants. A bewildering array of lush greenery covers the ground, climbs upon the trunks, and clothes the branches. The ground is completely hidden by ferns of myriad forms—some not recognizable as such to the uninitiated—heliconias with their bright flowers, and aroids of strange shapes, all of them shade-loving plants that can grow in the dimly lit forest interior. Trunks and branches are covered with mosses and moisture-loving lichens in shades of silver, pink, orange, green, and black. Great vines rooted in the ground—philodendrons among them—climb into the trees, where they disappear from view into the dense branches overhead. Roots from plants perched up above dangle down to the ground like ropes.

Plants that need more light and therefore cannot survive in the understory perch on branches in the upper levels, sometimes so thickly that it is difficult to make out the foliage of the trees themselves. Most of these species are invisible from the ground below, unless you view them from the edge of the forest. Here are orchids, bromeliads, cacti, begonias, anthuriums, gesneriads, huckleberries, and peperomias. They form an aerial garden whose roots never reach the ground and do not need to, for they have developed methods of living independently of ground water. They are epiphytes, *epi* meaning upon, and *phyte* meaning plant. They are not parasitic—their roots do not penetrate their hosts and they get no nourishment from them. Their roots, spongy-coated in some kinds, cling to the perch, wander through the mosses and lichens, intertwine with the roots of other plants, and sometimes hang free in the air. These roots collect water when it falls, and the plants store it in thickened stems and leaves. In most areas they are not subject to lack of water for any length of time, but in some regions there are definite wet and dry seasons, or dry periods between rains, when the plants draw on the water they store. Those that dwell in the upper

A tree trunk in a moist tropical forest, covered with mosses and lichens in which young bromeliads and orchids have taken root. (*Henry T. Northen*)

levels of the forest are more exposed to drying than those in the understory.

In many ways epiphytes have much in common with desert plants. Not only do they have thickened stems and leaves, but they are coated with a heavy layer of wax. Epiphytic cacti are but other species of the same kinds that inhabit deserts, and they are quite naturally equipped to collect and store water. Bromeliads hold water in a vase formed by a rosette of stiff, overlapping leaves. Their roots serve only as anchors, while the water is absorbed through thin membranes at the leaf base. Nourishment is brought to the epiphytes from collections of humus material lodged along the branches—dead leaves and insects, bird and animal droppings—dissolved out by the rains and fog.

In the trees and their aerial gardens lurk centipedes and scorpions; ants, bees, and wasps build nests; earthworms and snails make their homes; frogs, lizards, snakes, and spiders wait for prey; larger animals such as monkeys and ocelots move from limb to limb. Brilliant birds, butterflies and moths, and enameled beetles rival the bright colors of the flowers.

Even a single bromeliad is a little menagerie unto itself. The pool of water in its vase is a tiny permanent pond, with a population much like that of a natural swamp. In it dwell algae, microscopic animals, and worms. Water insects skitter on its surface, and larvae of insects develop from eggs laid there. Small frogs spend their entire life cycle—from egg to tadpole to adult—in the same plant. The larger bromeliads hold as much as a gallon of water. This aerial swamp makes it quite difficult to control the local mosquito population.

Few epiphytes harm the trees upon which they live, except that their weight occasionally causes a branch to break. One, the strangling fig, does eventually kill its host, not by stealing nutrients from it but by choking it. The seed of a fig germinates on the branch of a tree and grows into a large plant whose roots grow downward on all sides of the trunk, ultimately reaching the soil. As decades go by, the roots fuse, encasing the trunk and preventing its further growth. Meanwhile, the crown of the fig shades out the crown of the tree. At last the tree dies, leaving no evidence of itself, and the fig stands alone on a pillar of fused roots that outwardly resembles a solid trunk.

A wet tropical forest, a great blanket of life, has a very intimate relationship with the soil. The tremendous quantities of vegetation break the torrential rain, use large amounts of water, and support a huge

The pool of water in this bromeliad is a tiny pond that provides a home for many aquatic plants and animals. (*Henry T. Northen*)

population of soil fungi. In undisturbed forests soil fertility is maintained by the tree cover and associated plants. The minerals are recycled before heavy rains can leach them from the soil. For example, fallen leaves are quickly decayed by soil microorganisms and the released minerals are rapidly absorbed by the vast network of roots.

When the ground in a rainforest is denuded, the rains soon leach out the nutrients, washing them down to layers too deep for plant use. The fungi diminish because their source of food is removed. When a forest is cut down, farm crops do well for only a short time because the soil rapidly becomes impoverished. Corn and cabbage cannot take the place of forest in protecting the vulnerable soil. Even squatters who burn off small bits for their gardens must move on after a few years. If rainforests are to be farmed, the crops might well be tree crops that resemble the original forest. For example, *Cordia allidora,* a valuable timber tree, might provide an upper story below which food palms would thrive; beneath the palms, cacao trees could be grown. The lowest layer might consist of shrubs grown for medicine or fruit. In cooler regions of the tropics coffee trees grow in the shade of the Madre de Cacao.

A sad result of cutting off the forests is that many varieties of plants and animals are destroyed. No one knows how many species have already been lost, many probably never known to science. In countries where the exploding population desperately needs more land in cultivation, more timber for building, more room for cities, and extended systems of roads, the human need is so critical that mere plants and animals seem unimportant. We who have lost so much of our forested land and who look back with remorse upon the reduction of wildlife see an opportunity to prevent it in other countries.

Some groups are making a valiant effort to save certain species. Orchid growers, for instance, are trying to collect the wild plants and bring them into private collections and botanical gardens where they may be increased by seed. Zoos are trying to save the threatened animals. Although they may be considered of no economic importance in this commercial age, they are treasures of nature which cannot be replaced. Biologists are alarmed at the destruction of the forest canopy during warfare—in Viet Nam, for instance—and the long-term significance the devastation holds for life in those countries.

Forests can do what man cannot: protect the land, regulate the flow of streams, and give shelter for wildlife. That they offer scenic beauty and cool shade is a secondary blessing, but one which we treasure. In managing the nation's forests, the government agencies strike a balance between the utilitarian—such as watershed protection and a source of timber—and the recreational, making it possible for us to enjoy our forests and use them, too.

During the settling of our country, forests had to be removed to make way for farms and cities. Vast areas of hardwoods in the East vanished without contributing a board or a piece of furniture. Areas were slashed and burned. Logging was destructive and wasteful. Three-fourths of the wood was left to rot, and only one-fourth went to the mill. During the early history of the lumber industry logged-over areas were left barren and fires were seldom challenged. No thought was given to the future because the supply seemed unending. The lumber industry was nomadic, starting in New England, moving to New York and Pennsylvania, then to the Lake states, and still later to the South and the West Coast. Only in recent times did the lumberman come to realize that he must achieve a balance between growth and harvest, that he must keep the forests flourishing while taking their yield.

Where the land is covered with trees, wind erosion and dust storms do not occur, and water erosion is much reduced. The trees break the impact of rain and their wetted leaves allow the water to drip gently to the ground. The forest floor is a huge sponge which absorbs and holds the water, retarding its flow and helping prevent floods. Nearly half of the nation's streamflow comes from woodland, and in the West, 90 percent originates in forested watersheds. Most streams that start in forested areas flow throughout the year. Snow melts more slowly in the forests; in the West it sometimes lasts as much as six weeks longer than in the open. Thus, during May and June, when the streams are overflowing from spring rains and snow melting from exposed ground, the snow stays in the forests, to melt later and keep the streams flowing during the summer. The forest is critically important, too, for recharging ground water and providing for a continuous supply of surface water for irrigation, wildlife, and domestic and industrial use.

Forests cannot prevent all floods, but floods invariably become worse when forest is removed. When deforestation denudes the slopes, erosion occurs, with consequent damage to farms, lowlands, and cities. Silt washes into the streams and chokes reservoirs. Mud slides and avalanches sweep down hills that have been scraped bare—a situation that plagues heavily populated areas of California where homes have been built on the sides of canyons. Dams help with flood control, too, but reservoirs filled with silt are of no use. Some dams built within recent years have had a very short life. A reservoir surrounded by barren hills is particularly susceptible to filling-in, and more so if the slopes have been recently deforested and there is not even grass or weeds to hold the soil.

Since forests provide us with many necessary products—lumber for buildings and furniture; fiber and pulp to make paper, rayon, and cellophane; rosin for the violinist's bow; turpentine; charcoal, and an array of other things—trees must be harvested. But the method of harvesting should be planned so as to give a sustained yield of high-quality timber, and should be based on the characteristics of the trees.

When the species of trees are intolerant of shade—that is, when the seedlings cannot grow in the shade of older trees but require full sun—a method called clear cutting is used. All of the trees from any one spot are removed at once. If natural reproduction does not take place, seedlings are planted. This system is necessary for Douglas fir, jack pine, lodgepole pine, and the southern pines, kinds that tend to form uniform stands of trees of the same age. In order to protect the land, the cutting is done by strips or blocks, arranged so that water will be held by the still-covered areas above and below those cut over. The cutting is done over a period of years so that young trees will be starting on some areas and approaching maturity on others. When the last mature trees have been cut, the cycle starts over again. When trees are strongly rooted and will not be blown over with the removal of surrounding ones, a few may be left here and there to provide seed to repopulate the area.

In forests of shade-tolerant trees, such as birch, beech, and maple, the selection system is used. Since seedlings can grow in the shade of older trees, such forests contain trees of all ages, from merchantable size down to seedlings. Loggers cut the larger ones, leaving those too small for later. Then they go through the forests after some years, cutting those that have grown up in the meantime.

Thus, there are always trees in the area, and seedlings—a new crop starting each year—eventually take the place of the older trees that are cut. Foresters compare this method to Nature's system. In a forest undisturbed by man the older trees die and give way to young ones. In the selection system the older trees are harvested and the wood is used instead of being allowed to go to waste. This has many advantages over clear cutting, and it is unfortunate that it cannot be used for species intolerant of shade. It assures reproduction and uniform protection of the soil, and does not drastically alter the beauty of the landscape.

chapter fifteen

Grasslands

GRASSLANDS have always been the carpet of mankind; they attracted him and he has occupied them since before the dawn of history. He first ventured into the open spaces after game, and thus became a hunter boldly following great herds of wild animals. He frugally learned to use every bit of an animal: the skins for clothing and shelter, the sinews to sew the skins together, the bones for tools. He varied his diet of meat with the fruits and roots of wild plants, gathering them as he went. Some ingenious men tamed useful animals so that they no longer had to depend on the fickleness of the hunt. They became nomads, guiding their flocks to grazing lands, moving on to new places as the forage became depleted. And as they traveled, they gathered and took with them roots and seeds.

Some time in the dim past, men learned that they could cultivate plants and thereby control a supply of vegetable food. Their first efforts undoubtedly followed the technique of planting a patch and leaving it to itself while they went on their seasonal hunts or roamed with their flocks, returning in time to harvest their crops. Later developed the more stable practice of living near the cultivated plants and portioning out the jobs of hunting or caring for flocks to certain members of the tribe. Thus—perhaps as long ago as 6000 B.C.—began the early agricultural civilizations, the kind of life that gave stability to a community.

It allowed time for the development of arts from the skills the people had been slowly learning, for specialization for those with particular talents, and for improvement of the crop plants.

Nearly all of the food crops grown today were developed by these early farmers. All of the cereals are grasses; in both the Old and New Worlds they became staples—barley, millet, oats, rice, and wheat in Eurasia, corn in the Americas. Wild grasses still provide forage for cattle, sheep, horses, and many wild animals.

The grass family includes 4,500 species and populates about a fifth of the globe's land surface. Their diversity permits one or more kinds to grow as companions to other plants in practically all habitats—in tundra, desert, forest, in marshes, and on ocean shores. Only rarely are grasses cultivated for the beauty of their flowers, but the pampas grass is an exception. Although the flowers are inconspicuous and lack fragrance, viewed from close up they are seen to be delicately and intricately formed.

Two remarkable giant grasses are sugar cane and bamboo. Sugar cane belongs to the genus *Saccharum,* from "saccharon," an old Greek name for sugar. There are a number of species in the genus; sugar cane itself is *Saccharum officinarum*. It has been known to man for so long, and has been carried by him so far, that its native country is not known. It was widely cultivated in India and China, taken from there to Europe by travelers after the crusades, and brought to the New World during the 1500's. Its pollen is rarely fertile and plantings are made from cuttings; new stalks come from buds along the jointed stems. Sugar cane has long since been unable to meet the world's demand for sugar and is supplemented now by sugar beets.

In contrast to other grasses, bamboos are woody. There are about 200 kinds, ranging from some a few inches high to others 120 feet tall and a foot in diameter. They grow natively in both the eastern and western hemispheres, with over 160 species in Asia, 70 in the Americas, and 5 in Africa—but oddly, none in Europe. They are particularly abundant in the tropics, from sea level to 15,000 feet in the Andes, but many also grow in temperate and cold climates, even up to 10,000 feet in the Himalayas. The Japanese and Chinese have long represented them in paintings, and their decorative use in conservatories and gardens is world wide.

At least one kind of bamboo is a pest. It runs rampant where tropical forests have been cut down, and its tough, vining stems form such dense thickets that a man can work all day with a machete and penetrate only a few feet.

The uses of the valuable kinds of bamboos are myriad. Their hollow, flexible stems are strong for their weight, best known formerly in our country (before the advent of fiberglass) as fly rods. Larger stems are used in tropical countries for posts and rafters of houses, with split sections forming the side walls. Long lengths are used for water pipes and drain pipes, and short sections for containers and musical instruments. Split bamboo is woven into baskets, screens, mats, and fans. Gourmets everywhere, and the Chinese, especially, relish the young shoots of small varieties cooked, pickled, or preserved in sugar.

When bamboos flower—they don't all bloom, and many do so only rarely—they produce feathery clusters along the upper branches. Lack of flowers sometimes makes it difficult to determine species, a most frustrating situation for botanists. Also, they rarely set seed. Some that do form seeds continue to do so year after year, but others are so exhausted by the process that the whole plant dies. They propagate themselves by underground runners, often at such a rate that controlling their spread is a time-consuming job in warm climates.

The grasslands are called the "breadbaskets" of the world, and with good reason: they have the most fertile of soils. Their yield depends on the amount of precipitation they receive relative to the temperature. While forests develop in humid regions and deserts in arid ones, grasslands occur where precipitation is intermediate, generally between 12 and 30 inches annually, with most coming as showers during the growing season. Less rain is needed in a cool climate, for evaporation is low, and at high altitudes and latitudes grasslands may develop where the precipitation is only 10 inches a year. In the hotter regions of Africa, 50 inches may be required.

As one travels westward, precipitation gradually decreases from the 30 inches that fall along the Mississippi River to about 12 in the rain shadow of the Rockies. Coincident with the decreasing amount of moisture is a change in the vegetation. The dense green canopy of trees of the Mississippi Valley gives way to a sea of grass that extends to the mountains. In the eastern part of the grassland belt, in Iowa, for instance, between 30 and 22 inches fall annually, and here once thrived the tall grasses—big bluestem, little bluestem, and Indian grass—along with many showy flowering plants. A century ago these supported herds of buffalo and a variety of small mammals and ground-dwelling birds and their predators—the wolves, coyotes, hawks, and eagles. Only remnants of this prairie now remain, for the fertile lands have long ago been cultivated; corn, sorghum, alfalfa, and wheat have replaced the prairie grasses.

Where the annual precipitation drops to 22 inches a year, the tall grass prairie blends into the short grass plains, with a transition zone where little bluestem grows along with such short kinds as grama grass and buffalo grass.

As the precipitation tapers off from there to 12 inches and the elevation increases, the short grass plains begin and continue on to the foot of the Rockies. Grama grass, buffalo grass, needle grass, and an abundance of wildflowers populate the area.

On the plains the vegetation is about the same as it was a century ago, but it is now the foundation of the ranching industry. Sheep and cattle graze where buffalo once roamed. Antelope, once distributed over the entire area, are now limited to what open spaces remain, such as areas in Wyoming where fences do not overly restrict them, and where towns are small and are strung along the highways. The old-style barbed-wire fences, with three or four lines of wire, do not bother them; they can go through or under (they are less likely to jump them). Now, however, ranchers are changing to sheep-tight fences of woven wire, which limit their freedom. Sometimes they live perfectly happily in fenced range along with the cattle, but continued restriction may spell their doom. Today, however, one may still have the thrill of seeing herds of antelope kicking up dust as they run over the foothills and disappear in the distance. They are the most graceful of animals and the swiftest on our continent. When alarmed, they flash a signal by raising the white hairs on their rump and emiting a musky odor, then they charge away at 60 miles an hour, legs churning, backs perfectly level. Yet they are curious, and will stand and watch a person on foot or a car that comes to a stop, and when satisfied they may stroll off or run as the mood dictates.

Before the white man came, extensive prairie dog "towns" were common on the short-grass plains. Some covered many square miles. Today they are in limited areas and each covers only a few acres. Such towns are dotted with mounds, and in the middle of each is an entrance to a burrow. In digging their holes the prairie dogs turn over the soil, disturbing the original vegetation and allowing an assortment of succulent weeds to come in, unwittingly furnishing the precise condi-

Big bluestem grass once covered the tall-grass prairies.
(*U. S. Soil Conservation Service. Photo by Hermann Postlethwaite*)

tions for the plants that best serve them as food. One of the largest remaining prairie dog towns is near Devil's Tower National Monument in Wyoming. As a visitor approaches, the little animals that had been busily feeding suddenly sit upright like motionless pillars. At a sharp, high-pitched call from one of the elders, they all scurry into their burrows, parents even cuffing the young to make them obey.

Hawks and eagles still soar over the short-grass plains, meadow larks nest in the grasses and sing their incredibly sweet song, and badgers and foxes prowl. Coyotes still howl, but the large white wolves no longer roam the high plains.

Elsewhere in the world tall-grass prairies and short-grass plains occur where climatic conditions are similar to those of mid-America. Tall grasses dominate the pampas of Argentina and nearby Uruguay and Brazil, the llanos of Venezuela, and extensive areas of European Russia. Short-grass plains cover the steppes of southeast Europe and adjoining Asia.

The savannas, the great grasslands that cover so much of Africa, are actually forests held in abeyance by grazing and by fires. They are dotted throughout with mushroom-shaped trees that grow singly or in groves. The grass is lush. Sometimes it is 8 to 10 feet high, so tall that elephants can walk through it unseen. Yet it does not form a sod like the grasses of our prairies and plains—a sod so tough that it prevents tree seedlings from getting a foothold. There is ample rainfall for trees, divided into wet and dry seasons that correspond essentially to our summer and winter. Magnificent herds of wild animals graze the savannas—gazelles, elands, zebras, impalas, elephants, springboks, and a myriad others—as do domestic livestock. Thus, although trees can grow, only those that escape being eaten or trampled survive to mature.

From time immemorial the people have burned off the dead grasses during the dry season. This is good culture, for it removes the dense mat of tough, unpalatable dry grass and allows new shoots to come forth and rebuild a stand of tender forage. But, also, it keeps tree seedlings from growing. Thus man and animals together have maintained the savannas.

Wherever man keeps domesticated animals, problems of overgrazing arise, and the same problems can also occur in nature. In Africa, during the dry season the wild animals congregate around waterholes and inevitably overgraze the surrounding vegetation. People have allowed their livestock to cause deterioration of the savannas in parts of East Africa, and erosion has followed. Where both wild and domesticated animals occupy the land, the savannas are not in as good condition as where wild animals dwell alone. Ironically, areas that harbor the tsetse fly remain in best condition of all. They are avoided by herdsmen because the tsetse fly spreads the trypanosomes that cause the dread nagana disease in cattle and sleeping sickness in man.

Man's mismanagement has changed grasslands into deserts in many areas of the world. The danger is always present in regions of scanty rainfall or periodic drought, and overgrazing is a chief cause. Sheep are harder on the forage than cattle, and goats are even worse than sheep. The goats consume everything in sight, even desert shrubs that come in when the grass is gone, and it is said that they can, in desperation, climb trees to get the foliage. Sometimes a people is driven to raising goats when no other domestic animals can survive—in Greece, for instance—and this of course intensifies the damage. Grazing animals compact the soil, and as it becomes less porous, water does not penetrate as easily and runs off instead. Reduction of the vegetation exposes the soil to the drying sun and evaporation of water is accelerated. Each factor compounds the evil.

Desolate areas now encompass sites where many people once lived. The impressive ruins of Roman cities throughout North Africa are mute witnesses, as are poverty-stricken regions of Iraq, Libya, Turkey, Greece, and the Holy Land—the entire circle of the once-verdant cradle of civilization. Former grasslands in the United States have also been stricken, for example, those of the Salt Lake Valley and parts of Texas. In the latter, mesquite now covers what were once grassy plains.

After the Civil War enormous herds of cattle moved northward and westward from Texas over the "Old Chisholm Trail" and other routes. Sheep were brought in, too, to compete with cattle on public lands. At that time legal rights and responsibilities were poorly defined, with the result that cattle- and sheep-men tried to get ahead of each other to strip the forage before the next herd or flock arrived. Range wars flared. With each passing year more animals grazed the range. From 1870 to 1890 cattle increased from 4,500,000 to 26,000,000; and between 1850 and 1890, sheep, from 500,000 to 20,000,000. The western range deteriorated and continued to do so until the 1930's.

Since the Thirties a more orderly use of the public range has prevailed. Ranchers are granted specific areas on which to graze their animals, paying a fee for each head, and the number is adjusted to the supply of forage. Ranchers are also learning how to manage their home acreage to better advantage. They have found that they achieve a higher return in meat and wool when the land is not overstocked. They avoid the tendency to concentrate animals on the best meadows and along streams, making use of poorer areas as they can. They let the grasses become well started in the spring before turning the livestock out. They achieve a more even distribution of animals by riding, herding, and fencing, by placing alluring salt blocks in spots not normally grazed, and by developing additional watering places. Reseeding of wasted areas is being successfully carried out.

Although some of our farmlands were once forested, the greater percentage have been developed from grasslands. Removal of the native vegetation in either case creates problems that involve us all, for soil is an irreplaceable treasure. The loss of topsoil through wind and water erosion has been and remains a major problem, one that must be solved to avert disaster. Of the 1,904 million acres of the United States, excluding Alaska, 510 million are in cultivation. However, land suitable for cultivation is diminishing through misuse and through diversion from farming. Each year about a million acres of choice farmland is put into highways and subdivisions. Meanwhile, much land is being farmed that would be better put to other use, and careless treatment is ruining further acreages.

The drought, dust, and despair from wind erosion during the early Thirties wakened the nation to the need for soil conservation. Farms in Colorado, Nebraska, Kansas, Oklahoma, and bordering states were struck by prolonged drought. The dry soil of the ploughed land was blown away in choking clouds that spread across the continent, darkening skies and reddening sunsets. Steinbeck in the *Grapes of Wrath* immortalized the plight of the farmers who migrated en masse to California, where they found no employment and lived in abject poverty. Wisdom of hindsight shows that much of this land should never have been farmed. The fact that it was fertile and inexpensive, and that with enough rain it gave good yields, tempted people to try it. But over the whole area droughts are a frequent occurrence. Farming is hazardous enough at best, and the farmer has enough trouble from periodic infestations of insects, and from unexpected freezes and hail storms, without begging trouble by farming where the odds are stacked against success.

When topsoil is exposed by having its cover removed, it can be washed or blown away in just a few years. In drought-prone areas it is best left with its native grasses to protect it and the land used for grazing. Even in areas with plenty of moisture, ploughing steep slopes allows the soil to erode away. It may be completely gone in twenty

Careless use of farmlands brings disaster and despair. Here, wind and water have eroded the soil, and gullies have formed. (*Soil Conservation Service*)

Wise use results in beautiful and productive farms. Here, strip cropping and contour farming is practiced and trees protect the steeper slopes. (*Soil Conservation Service. Photo by Lathrop*)

years. Steep slopes should be left in grass or forest. About 14 million acres in the Great Plains which are currently under cultivation should be returned to grass.

Gentle slopes in regions of adequate moisture can be farmed without excessive risk if certain practices are followed: the land should be ploughed along the contours so that the furrows form small dams to hold the water instead of allowing it to run downhill and form gullies. Terracing is also effective in reducing erosion, as is lister ploughing. In the latter, adjacent furrows are thrown toward each other, leaving between them a ditch about 6 inches deep in which the crop is planted. Strip cropping is remarkably effective in reducing erosion and creates a beautiful and charming landscape as well. Bands of cover crops such as alfalfa or clover alternate with bands of row crops such as corn, potatoes, or cotton, or with broadcast (scattered) ones such as wheat or oats. The soil washed off the row crops is captured by the cover crops. Periodically the plants are rotated, which helps to maintain soil fertility and to lessen the damage by insects and disease.

Maintaining soil fertility is a prime goal. Under natural conditions there is a continuous recycling of nutrients, and the fertility remains constant. With agriculture, however, soil minerals go to market with the crop. A ton of wheat takes away about 40 pounds of nitrogen, 8 pounds of phosphorus, and 9 pounds of potassium. Two steers sent to market represent a soil depletion of 54 pounds of nitrogen, 15 of phosphorus, and 3 pounds of potassium. Such rates of removal must be compensated for by adding fertilizer to the soil.

The Soil Conservation Service in the Department of Agriculture, established in 1935 following the dust bowl disaster, works with farmers to reduce the wastage of soil and to maintain its fertility. Today it functions through 2,700 Soil Conservation Districts to develop plans for soil and water conservation on crop and pasture lands and on woodlands.

Although we rely on the grasslands and farmlands to feed us, countless millions of us who never see a ranch or a farm feel the need of grass in our surroundings. We carefully nurture and manicure our lawns, we create velvet-smooth golf courses, we plant green strips between highway lanes, and we maintain parks for our children (though here there are too many "Keep off the grass" signs). More and more do we beautify public buildings with grass, even though it may be only a narrow border along the sidewalk. Grass continues to be for us, as it was for our ancestors, the carpet of mankind.

chapter sixteen

Tundras and Deserts

NOT all plants are destined to grow where the climate is congenial. Some, the plants of the tundra, must live where winter's cold and fierce winds prevail for nine to ten months of the year. Others, the plants of the desert, survive where drought is frequent and prolonged. In both places the plants have become adapted to their harsh environment by means of ingenious modifications.

The tundra stretches in a circumpolar belt across northern America, Europe, and Asia, and caps the high mountains of temperate zones below the line of permanent snow. In a setting of austerity and often of bold and magnificent scenery, the plants are of Lilliputian size. At 11,000 feet in the Rockies, the trees are dwarfed, twisted, and gnarled, and the flowering plants are matlike perennials barely a few inches tall. As one goes farther and farther north, tundra is found at ever lower elevations, until in the Arctic it reaches sea level. In all tundras the summers are short. The plants must hurry through their life cycle, producing leaves, flowers, and fruits often in two months' time.

You will see similar vegetation on the high mountains of all temperate climates, whether you visit the Rockies, the Cascades, the Alps, or the Himalayas. In fact, the term *alpine* is used to describe them all. Few spots are more colorful than the alpine tundra where the ground is a mosaic of color, of pink *Silene* and lavender *Erigeron,* of yellow

Potentilla, and blue mountain forget-me-not, of white and blue phlox. Everyone knows edelweiss, beloved flower of the Alps, about which songs and poems have been written. The little alpine plants cling close to the ground, spreading out in mats or cushions for protection against the wind, their leaves reduced in size and covered with fuzzy hairs to protect them against drying. They spend most of the year under a blanket of snow, and as the snow melts and the sun warms them, they burst into bloom. Close to the permanent snow, where the edges of drifts melt back as late as August, you may find tiny willows pushing forth their "pussies," with barely enough time to mature seed before snow covers them again.

Here the ptarmigan moves with no fear of man. You may see it with white feathers among the brown in preparation for winter. The pika is busy drying its little stacks of hay on rocks before storing it. Sleek marmots whistle among rocky outcroppings.

In the Arctic tundra, the environment is even more severe than in the alpine, because the ground is permanently frozen. Only a shallow upper layer thaws during the summer, and the soil is often waterlogged for lack of drainage. In this tundra, also called the Arctic Barren Grounds, a variety of grasses, sedges, dwarf shrubs, flowering plants, and the lichen known as "reindeer moss" survive the winter and burst into activity during the long days of the brief summer. Many birds and animals migrate to the Barrens as spring flows northward, and then as summer wanes they return to warmer regions. A high percentage of America's waterfowl nest and feed in this vast area. Animals that have learned to cope with the Arctic environment make it their permanent home—the caribou, Arctic fox, musk ox, and polar bear. During the summer they put on a layer of fat, which tides them over the winter when foraging is difficult.

Among men, only the Eskimos developed the cultural skills to be self-sufficient in the Arctic. The coastal tribes secured life's necessities from the abundant marine life. The caribou hunters relied on the resources of the land and, though they hunted other animals as well, relied mainly on the caribou for food, clothing, and other necessities. Contact with the modern world has brought about a mixed condition, for while seeking some of our comforts, they are losing old skills, and poverty plagues their present state.

The problems associated with living in such a severe environment have left the tundra itself little spoiled by man, although some of "civilized" man's activities once threatened all life in the area. During the period of competitive testing of nuclear weapons, enormous amounts of radioactive fallout contaminated the plants of the tundra. Dangerous minerals such as Caesium 137 accumulated in the reindeer moss. The flesh of caribou, whose major source of food is this lichen, became radioactive, and consequently, so did the bodies of the Eskimos who ate the meat. No serious immediate illnesses have shown up so far, but scientists are searching for possible long-range effects, including genetic damage. Here is an example of how air pollution can affect distant and innocent people.

The pioneers suffered great hardship as they struggled across the Great American Desert. Water was scarce, and often bitter and unpalatable. Their disappointment at some creeks led to the names that now fascinate us on maps—Bad Water, Bitter Creek, No Water Creek, Alkali Creek, Poison Water. Gratitude at finding a river they could follow for a while is reflected in the name Sweetwater. Even today one must exercise great caution before taking off on a secondary road or trail—a full tank of gas, a reliable car, and an ample water supply are necessities. Most deserts are lonesome areas, still in pristine condition, with few human inhabitants and none of the facilities offered along well-traveled roads. There are places where one can drive a hundred miles without seeing a house or meeting another car. Nevertheless, the solitude is appealing to the venturesome, the beauty is alluring, and the wonder of the desert holds fascination.

Throughout the Great American Desert, and deserts elsewhere, precipitation is scanty, generally less than 10 inches annually. In many areas it is unpredictable. Sometimes a third of the annual amount may come in one cloudburst, producing a flash flood that fills arroyos with rushing torrents and washes yet another layer off eroded slopes, carving weird and fantastic land forms. In 1968 a wall of water seven miles wide and several feet high washed out a town in Utah whose residents didn't even know it had rained to the north, where the flood originated. A pair of zoology students collecting there but a few hours before were shocked to learn the next day that the town where they had stayed no longer existed. Although such cloudbursts may cause some casualties and are a threat to human beings, they are beneficial in the long run to plant and animal life.

Climates change, and the deserts we see today were not always desiccated lands. Many were once forested and were inhabited by a wealth of animal life. The petrified trunks of trees that lived millions

The land in Petrified Forest National Park, Arizona, once supported trees, the size of which can be seen in their petrified remains. (*Henry T. Northen*)

of years ago are now eroding out of the sands, beautiful in cross section, where the minerals deposited by water gave them rainbow colors and filled crevices with crystals. Petrified Forest National Park in Arizona preserves such an area. Some lands were covered with advancing and receding inland seas, being entirely under water at some times and swampy regions at others. The bones of ancient marine and land animals are mute testimony to their past history. Uplifting since then has allowed these lands to be worn away, and the strange and beautiful results can be seen in Bryce and Zion National Parks, the Badlands of South Dakota, and in many other places.

In spite of the adverse environment a desert offers, the desert inhabitants are well adapted to it. Some of the plants simply escape drought; others withstand it in special ways.

Those that escape it are the annuals that color the desert magnificently after heavy rains. In the interval between rains they survive as seeds which are extremely resistant to drought. After heavy rains, the seeds germinate, the plants develop, come into flower, and ripen and scatter their seeds—all in a matter of a few weeks. Because they complete their life cycle in a short time, appearing and disappearing quickly, they are called ephemerals. A possible record is held by an annual in Timbuctu, *Boerhaavia repens,* which goes from seed germination to seed ripening in just eight days. Summer annuals of the deserts of our South and West are *Aristida adscensions, Boerhaavia spicata,* and species of *Euphorbia, Mullugo, Portulaca,* and *Pectis.* Spring annuals are *Eriophyllum wallacei, Gilia aurea, Hemacladus longiflorus,* and *Plantago spinulosa.*

Cacti and various succulents resist drought by storing water in fleshy stems or leaves, providing themselves with a reservoir from which to draw during spells between rains. Most notable is the saguaro cactus of the Sonoran Desert. Its water-storing stem is a fluted column which contracts and expands like an accordion, shrinking as the water is used, and expanding when the supply is renewed. Its roots radiate 50 feet from the trunk and, except for one anchoring taproot, spread just below the soil surface. After a rain the roots avidly absorb water and move it to the gigantic column, which can hold as much as a ton—about 1,000 quarts. In addition to water-storing stems, the cacti have leaves that are reduced to mere spines, a characteristic that cuts down water loss. Food-making has been taken over by the stem, and the spines' sole function is that of protection.

Some very clever habits are shown by leafy drought-resisting plants. Their leaves for the most part are quite small, and in many species both leaves and stems are coated with a thick layer of wax. Most have but few breathing pores (stomata) through which water escapes, and these are sunken in pits or grooves where the air is protected from the wind and where it remains relatively humid. In other species the leaves are covered with downy hairs that reflect light and at the same time lessen evaporation of water from the surfaces.

Strangest of all are the kinds that evade drought just as deciduous trees evade the cold. They produce a full array of leaves when water is available and drop them during dry periods. The paloverde has minute leaves only one-eighth of an inch long, which it soon sheds, leaving the leafless green stem to carry on photosynthesis. The leaves of the ocotillo emerge from buds after a shower and are shed when drought comes, and this may happen as often as six times a year.

Desert animals as well as plants are adapted to dry conditions both by habit and by their physiology. Some escape the heat by remaining in cool burrows during the day and by congregating around the few permanent streams and waterholes. Some hibernate or estivate during dry spells and come out only when it rains. Such are the desert or spade-foot toads which burrow into the soil, as well as some of the ground squirrels. Many animals seldom find water to drink but get their moisture from the plants they eat—and from the dew that collects upon them. They minimize water loss by excreting a highly concentrated urine. Numerous kinds of animals can tolerate the saline water found in alkali lakes and puddles, excreting the extra salt. The kangaroo rat is the most remarkable example of drought tolerance. It doesn't even have to eat succulent plants, but can live on dry seeds alone. All living things manufacture a little water within their systems during the process of respiration, but for most it is entirely inadequate to maintain life. However, this water is all the kangaroo rat gets, and it can live on this trace. Its unusually strong kidneys concentrate its wastes so completely as to conserve almost all of the body moisture.

The Great American Desert is a term that covers the deserts of our West, which also extend into Mexico and which comprise two distinct types: the Great Basin desert, where rigorous winters as well as hot summers prevail, and the much warmer Southwestern deserts.

The kangaroo rat of the desert never drinks water.
(*Interior—Sport Fisheries and Wildlife. Photo by E. R. Kalmbach*)

The Great Basin desert covers most of Utah and Nevada, and extensive areas of Wyoming, Colorado, Idaho, and Oregon. Here precipitation is less than 12 inches annually, fairly well distributed throughout the year. Often this area is known as the Sagebrush Desert, because sagebrush is the dominant and most conspicuous plant. However, other kinds flourish as well: cacti, many grasses, and flowers such as Indian paintbrush, larkspur, locoweed, phlox, penstemons of great variety, sand lilies and a myriad others. A few are poisonous to livestock—larkspur and the infamous locoweed (a legume) among them—but most are nutritious and support a cattle and sheep industry.

To appreciate the great variety of animals associated with sagebrush, one must leave the car and walk through the area. Here and there he may flush sage grouse, meadow larks, and other birds; jack rabbits, cottontails, and badgers. In the distance he may see coyotes and antelope, and soaring overhead, hawks and eagles.

Freshwater lakes occur in the Sagebrush Desert, but briney lakes and salty swamps appear in places that have no external drainage. Some plants can thrive in the very salty soil—greasewood, shadscale, winterfat, and saltgrass, for example—and around the borders of the alkaline lakes grows glasswort (*Salicornia*). In some respects, the life in the briney water is similar to that in the ocean. If you take home some of the dried salt from the rim of a lake and put it into a dish of water, brine shrimp may hatch out from eggs which were laid months before, and which were only waiting for a rain to fill the lake to its borders again.

The Southwestern Desert extends from Nevada south into Mexico, with three major regions: the Chihuahuan, the Mohave, and the Sonoran. While cactus and creosote bush thrive over the area in general, other plants are more restricted in their distribution, some so characteristic that they have become indicators of the three divisions, emblems of their respective territory.

The Chihuahuan Desert takes its name from the Mexican state in which most of it is located, from whence it extends northward into Arizona and Texas. Agaves are its outstanding plants—there are so many species of them that many may not yet have been discovered. Others remain nameless because they are rarely found in bloom. Their sword-shaped leaves, tipped with a spine and also often edged with spines, form stiff bluish or silvery bouquets. From the center rises the

The century plant, *Agave americana,* one of many agaves that live in the Chihuahuan Desert. (*U.S. Forest Service. Photo by Bluford W. Muir*)

flowering spike, in some species as thick as a man's arm and more than 20 feet tall, bearing hundreds of lilylike flowers of yellow, purple, or brown.

It takes years for an agave to store enough food to support the huge flower stalk. Twelve to fifteen years are required for the flowering of the largest mescals; thus, they have been given the name of century plant. In some species, the parent plant dies after its magnificent, one-time production. In others, however, new plants grow from the old roots.

Still large, although smaller than the mescals, are the lechuguillas, the true emblem of the Chihuahuan Desert. In the spring their 12-foot flower spikes rise across the desert bearing plumes of color. Since prehistoric times, people have obtained fibers for cloth from this and from the maguey. It is said that the maguey gives them not only the cloth but the needle and thread to sew it, for when the leaves are macerated to obtain the fibers, the "thread" attached to the spine at the leaf tip forms an already threaded needle. Even today, on back roads in Mexico the people still spread the leaves where cart wheels and burro hooves will shred out the fibers, and the juicy tissues dry away in the sun.

Alcoholic beverages are made from various species of agave, supposedly different drinks from different kinds—pulque, mescal, and tequilla. They are obtained from the sweet sap of the flower stalk, cut off when young and then fermented.

The Mohave Desert occurs in southern California and Nevada. The driest, hottest part is typified by Death Valley National Monument, the lowest point on the continent, 279.6 feet below sea level at Badwater, where the total precipitation for the year is less than 2 inches. The temperature reaches 134° F.—heat which would sear nonadaptable plants. Yet many plants and animals are at home in this difficult environment, kinds which do not thrive in more moderate climates. Six hundred plant species dwell in Death Valley National Monument, and 230 birds have been reported. Many of the latter are transitory migrants, but at least 14 species nest on the valley floor.

Until a rare rain beats down, many plants are inactive and others, the ephemerals, invisible. The coming of rain rejuvenates the shrubs and cacti and brings about germination of the ephemerals, so that the ground becomes carpeted with the yellow, rose, and white of the satiny cactus blooms and the yellow of poppies and sunflowers.

The Joshua tree is the emblem of the moister parts of the Mohave Desert. (*National Park Service, Interior. Photo by Grant*)

The Joshua tree is the emblem of those parts of the Mohave where the precipitation reaches 8 to 10 inches a year. Its tall, weird forms reaching as high as 40 feet dominate the landscape. It grows nowhere else in the world. It is a species of yucca, and is called the tree yucca as well as the giant Joshua. Unlike other yuccas, it has a trunk and branches. Its spine-tipped leaves grow in bunches at the ends of the branches, where tight clusters of greenish-white flowers appear during the spring. The porous wood furnishes nesting places for birds, lizards, woodrats, and insects, and Indians use its red roots for designs in basketry.

The Sonoran Desert, perhaps the most fascinating of all, covers southwestern Arizona and the adjacent section of Mexico, most of Baja California, and the southeast corner of California. It takes its name from the Mexican state of Sonora. Here the stark shapes of the giant saguaro cactus, emblem of the Sonoran Desert, are outlined against the sky, giving pause for wonder at their strange and grotesque forms. Where the Sonoran and Mohave Deserts border each other, the saguaro and the Joshua tree grow together, confounding the human mind at the strange whims of nature.

The saguaro cactus grows to 50 feet tall and can weigh 12 tons. It is a member of the genus *Carnegiea,* related to the night-blooming cereus most people know. It, too, opens its 6-inch waxy white blooms at night, and closes them the next afternoon. Because it stores such great amounts of water, it is not dependent upon rain for flowering and can bloom during extended dry spells. Long lived, it is believed to attain a maximum of 200 years. Unfortunately, the saguaro is disappearing from all but protected lands, for many areas are now being irrigated and placed under cultivation; but they are preserved in Saguaro National Monument in Arizona, and can also be seen in Organ Pipe Cactus National Monument.

Numerous shrubs, both evergreen and deciduous, other kinds of cacti, and many ephemerals share the region with the saguaro. The precipitation comes during two periods, summer and winter, with exteme dryness in between. As in all deserts, the ground is covered with an unequaled wildflower display after a period of rain.

The life of many animals centers around the saguaro. Lizards, spiders, and moths inhabit the trunk, as do woodpeckers. The woodpeckers excavate their nesting holes sometime prior to the nesting season so that when the time arrives, the holes will be dry for the eggs

The saguaro cactus is typical of the Sonoran Desert. *(National Park Service, Interior. Photo by George Olin)*

to be laid and the young to be reared. The nests are then abandoned and the holes occupied by other birds—elf owls and flycatchers. When these move out, the holes offer shelter for mice, rats, snakes, or lizards.

Perhaps the tundra and desert should be considered inhospitable only by man's standards. The plants and animals that dwell there are perfectly at home; they have learned to live with what we consider formidable drawbacks. Through eons of time they have evolved to live in tune with their environment and with each other—a challenge that man is only now, late in his evolution, beginning to face.

chapter seventeen

Succession and Competition

As living things have migrated over the face of the earth, plants have always preceded animals, making the land hospitable for them and offering food and shelter. Plants have been the pioneers, opening up new areas for those who would come after them, and of these, the simple plants have preceded the more complicated ones.

Plants are still colonizing the earth today. Always in the vanguard are the lichens. They spread over expanses of bare rock, such as recently cooled lava fields or mountainsides denuded by landslides or blasted for highways. They are the last plants seen as one approaches the pole or ascends the summits of the world's highest mountains.

Even among the lichens there are some that are more hardy than others. The crustose ones, silver, orange, and black, are the first to cover the rock, adhering so tightly that they can scarcely be removed, either by animals or by the elements. As they dissolve minerals from the rock, they slowly crumble its surface. They catch dust blown by the wind and washed down by rain and snow. As bits of lichen die, bacteria come in to feed on the dead material, releasing minerals for reuse by the lichens and furnishing humus to the minute amounts of soil.

Once the area has been homesteaded, other types of lichens invade, and thus begins the competition for living room that never ceases in the plant world. The newcomers are the gray and blue-green foliose kinds, taller than the crustose ones in a Lilliputian world where height is measured in fractions of inches. The two share the area for a while, but gradually the larger crowd out the smaller. The foliose lichens continue to build the soil and eventually make the area inviting to new invaders, the mosses.

Moss spores are brought in by the wind. It takes a hardy kind to survive the still rather harsh conditions, where the soil is less than a quarter of an inch deep and can hold but little water, but the rugged blackish-green mosses can do it. They, too, add to the soil, and bacteria feeding on the accumulation of dead material enrich it.

After centuries or millenniums, the soil becomes deep enough to hold sufficient water for flowering plants, whose seeds, wind-borne or carried by birds or animals, now take root. The low-growing mosses are doomed, for they are no match for the newcomers. A colorful garden soon covers the soil, made up of grasses, bluebells, sedums, skullcap, cinquefoil, and a myriad others. The parade goes on as shrubs follow these—roses, currents, and gooseberries.

Animals follow the migrating plants, moving in as soon as the area is suitable for them. Microscopic soil dwellers come first, then worms and burrowing insects, then flying insects, toads, lizards, ground-nesting birds, and small mammals, followed by their predators. Each contributes to the community. They enrich the soil with their waste matter; the insects pollinate the flowers; the animals distribute the seeds.

Trees make the next move, but whether they can take over the area depends on the amount of moisture available. Trees generally require more water than grasses and shrubs. Seeds may germinate and young trees start, but if the competition for water is too great, they will lose out to the plants already present. With sufficient water, however, they will thrive, crowd out the shrubs, and eventually develop into a forest.

The succession continues even among the trees. Over the greater part of the United States, the sun-loving pines are the first to grow, the particular species varying from region to region with the climate—lodgepole pine in the Mountain West, jack pine in the Lake states, and loblolly pine in the South. In the northern forests the first trees are aspen, balsam, and poplar. Pines are very gregarious, usually growing in such thick stands that no other light-demanding plants can compete with them. Ironically, they crowd out their own seedlings, which cannot develop in the dense shade of the older trees. However, seeds of shade-tolerant plants, some brought by the wind, some by animals, can thrive under the canopy. Whenever a pine tree dies from accident or disease, its place is taken by one of the tolerant kinds that has been biding its time in the understory. Gradually the pines are replaced. In the Rocky Mountains they give way to spruce and fir; in the Carolinas, to oak and hickory; and in parts of New England, to maple and beech. And these trees remain unless they are destroyed by man or by some natural catastrophe.

The final vegetation in any one place is called the climax. It is governed by the climate, of course, and by the annual moisture. In the far North in some places succession stops with the lichens, the so-called reindeer moss. Where it stops with the grasses and flowering herbs, you find prairies, plains, veldts, pampas—the grasslands of the world. Where it ends with trees, you have a climax of forest.

The climax forest is self-perpetuating. It allows seedlings of its own kind to grow in the understory, along with shade-tolerant shrubs and flowering plants. The forest protects its soil from erosion and continues to build and enrich it. The streams run with clear, silt-free water. Fungi bring about decay of fallen leaves and branches, keeping the trash cleaned up and returning essential minerals to the soil. Thus, raw materials are used over and over, and the topsoil will be as fertile a thousand years from now as it is today. Mushrooms and moisture-loving kinds of mosses and lichens dwell alongside the more showy plants. The forest is also the home of a characteristic assemblage of animals and birds. Each living thing finds its niche, and each is important to the community.

The next time you visit a climax forest, contemplate the events that brought it about. The harmony that prevails is the end of a long history of strife and invasion. Each population was upset by invaders, and it was not until kinds arrived who could live together that peace or permanence came about. In the absence of a major calamity, that permanence should persist. Man is most likely to be the cause of disaster, although nature can produce catastrophe too—fire caused by lightning or volcanic eruption, for example. It behooves us, the inheritors of the climax forests, to keep some of them safe from destruction. If we destroy them they will not be built again in our lifetime, or for many generations after us.

Even though the climax community is permanent, there is still competition between individual plants for water, light, and minerals. The animal occupants vie with members of their own species for territory, and large animals prey upon smaller ones. But there is cooperation, too. Shade-loving flowers and seedlings rely on the trees above them, and both depend on the soil-dwelling animals to aerate the earth and release minerals for reuse. The animals depend on the forest for food and shelter and would become destitute without its protection.

Plant succession takes place in lakes and ponds as well as on land. No lake is permanent. When a lake is young, its bottom is sandy or rocky and the water is clear. The pioneers who move into the lake are few at first, consisting of floating and submerged algae, bacteria, and microscopic animals. Slowly the population builds up until there is food for insect larvae, crustaceans, snails, and those fishes that require fresh, open water—in cool areas, trout, and in warmer ones, bass. By now the lake teems with life, and there is death along with birth. As the dead settle to the bottom, their remains cover it with a muddy

A lake in Alaska in the process of being filled in. The once clean sandy bottom becomes covered with dead plant and animal remains, building a layer of soil. Submerged plants take root, and then emerging plants come in around the edges. The shoreline becomes swampy and marsh plants grow there. As soil is built farther out into the lake, sedges and then flowering plants, and finally trees are able to grow. The lake will eventually become entirely filled in.
(*Rebecca T. Northen*)

layer, causing the lake to become shallower with each passing year. When it has become shallow enough—less than 20 feet—submerged plants which grow entirely under water take root in the mud: *Elodea,* pondweeds (*Potamogeton*), submerged buttercups, and various kinds of algae. Their presence slows down the flow of currents through the lake so that silt carried by incoming streams settles to the bottom. Finally, plants cover the entire lake floor, filling it with a tangled aquatic garden so dense that boating and fishing are sometimes difficult.

At the same time, plants encroach along the shore, building the soil ever farther into the water. Dead branches, trees, and leaves fall into it, increasing the material on the bottom. Eventually the shore becomes swampy, and cattails, bulrushes, and reeds appear. In the shallow water at depths of 2 to 5 feet, water lilies and smartweed take root. They are joined by floating plants such as duckweed, and together they may cover much of the lake's surface. Trout and bass now give way to fish that can thrive in the more sluggish water and heavier vegetation, and frogs, newts, and diving spiders move in.

As time goes on the rate of filling is accelerated (the larger the lake the longer it takes). Eventually it becomes a marsh. The cattails, reeds, and bulrushes multiply more rapidly now, and soon take over completely. After a while they give way to what is called a sedge meadow, with sedges, mints, marsh marigold, iris, and bell flowers, a beautiful array and lovely during their flowering season. But they, too, are transitory. As the soil builds up, grasses and shrubs come in, and may be the climax or may give way in turn to trees.

Here again, since the actual species are dictated by the climate, the climax will be different in different regions. One cannot always casually tell whether it developed on an area that was originally bare rock or upon a filled-in lake, unless the contour of the land makes it clear. Otherwise, the only way to find out is to take deep cores from the soil, for an old lake bed will often reveal accumulations of pollen from past ages along with other vegetable matter, incompletely decayed as it collected in the bottom.

When the plant cover is removed, either artificially or by a catastrophe of nature, the soil can be blown or washed away, often down to the bare rock. When forests are burned or cut on steep slopes, floods wash the soil into the valley below. Dust bowls have developed where land has been overgrazed or put to crops in areas of recurrent drought.

When the soil is entirely gone, succession has to start all over again from the bare rock.

In areas denuded of plants but where the soil remains fairly undisturbed, plants can come in again quite peacefully. This is called secondary succession. A study was made of abandoned cotton fields in North Carolina where the sequence could be seen in all stages. First came weeds—crabgrass, horseweed, asters, ragweed, and broom sedge—from seeds brought in by wind, birds, mammals, and water. Within a few years, pine seedlings appeared amidst the assemblage, and gradually they shaded out the first plants. After 75 years the pines had grown quite tall, and seedlings of hardwoods were developing in the understory. After many decades, the hardwoods would replace the pines, and the climax forest would be one of oak and hickory.

You can probably see secondary succession in areas where farms have been abandoned, unless they have sprouted with housing developments. On a small scale you can see it in your garden. If you neglect the flower beds, weeds soon push out the desirable plants. An old home left uncared-for slowly becomes hidden by rank growth of weeds and shrubs. Some of the latter may be wild, and here and there young trees of species not planted by the former occupants may make their appearance.

People in London after World War II were astonished to see flowering plants lining the bomb craters. There was much speculation as to how they got there, for some were kinds never before seen by Londoners. Some undoubtedly sprang from seeds long buried which were able to germinate and grow when the soil was turned over. Others were kinds that can grow only in disturbed soil; lacking the vigor to compete with other plants, they grow best where the plant cover has been removed. Their seeds may be brought in every year, and seedlings may start unseen, only to perish. But when they have a chance to grow, they do so promptly. Such are the fireweeds, species of *Epilobium,* whose seed is wind-borne. Quite spectacular with rose or orange flowers, they are common to both Europe and North America. They suddenly appear in burned-over areas, but also grow along roadways and on any spot where the vegetation is thin.

One of the loveliest sights of an Alaskan summer is the fireweed, which sweeps over the hillsides in unbroken color amid the blackened remains of burned "Taiga" forest (forest of little black spruce). There, where the water of the thawed surface soil produces a near bog above the permafrost, the fireweed has the wet conditions it likes best, and it grows in riotous profusion seldom equaled elsewhere.

When part of a forest is cut or burned and little or no erosion takes place, it will regenerate itself in time. The weeds and shrubs that come in help to stabilize the soil. They are accompanied by mice, rabbits, and other small mammals, and are browsed by deer, so that they are actually beneficial to wildlife in some respects. As you drive through the countryside you may see the stands of young pines that soon follow, stands of trees all the same size. If the region is left alone, the trees which formed the original climax will come in. However, the pines are desirable for lumber, often more so than the climax trees, and lumbermen may thin out and carefully nurture them until they can be logged over, to allow another crop to develop in its turn. Lumber companies now practice reseeding after cutting, and where this is carried on you can see tree farms in all stages of growth.

The wildlife picture changes considerably when a climax forest is destroyed. A study of Wells Gray Park in British Columbia tells one instance of such a change. This was a primeval area consisting of high mountains, foothills, and valleys with streams. A cedar-hemlock forest covered the valley floor and lower foothills, where the damp climate also allowed a growth of moisture-loving lichens. Spruce-fir forests occupied the north-facing slopes of the higher elevations, and Douglas fir and grassy meadows, the south-facing ones. Many kinds of animals lived in the varied environment, but in small populations. Large herds of caribou came to winter among the hemlocks, where they fed on the abundant lichens growing along fallen trees and on the tree trunks.

Between 1926 and 1940 almost 800 square miles were burned by fires so intense that they even destroyed the humus of the forest floor. Before the fire, there were mountain goats, mule deer, mountain lions, coyotes, black and grizzly bears, wolverines, martens, beavers, and the annual herds of caribou. After the fire, the caribou disappeared owing to the loss of their winter food. Also gone were the grizzly bears, wolverines, martens, and mountain goats. As the valley floor became overgrown with shrubs, chiefly willow, the deer, black bear, and beaver came back, as did their predators, mountain lions, and coyotes. Moose discovered the new grazing grounds and invaded in large numbers, and timber wolves came to prey upon them. As always, mice and ground squirrels arrived with the weeds that now covered the burned-over slopes.

Mono Dam and Reservoir was built in 1935 to catch soil and debris washing down from slopes burned by a fire and to prevent filling of the larger Gibraltar Reservoir. Heavy rains the following two years washed down enough debris to completely fill this reservoir and partially fill Gibraltar. This photo was taken in 1938. (*U.S. Forest Service. Photo by F. E. Dunham*)

This photo, taken in 1949, shows how quickly plants moved in to grow in the moist and fertile silt—first, grasses and weeds, and then shrubs and trees. This is an example of secondary succession. (*U.S. Forest Service. Photo by W. I. Hutchinson*)

You might say that it is not so bad when a plant community changes; some animals move out and others come in. Indeed, it would not be bad if there were always equivalent homes for those who move away, but the sorry fact is that homes cannot always be found. Huge numbers of animals have starved in the process of being driven out of the wild areas that are their natural habitats. Species that depend on wilderness areas of various types are approaching extinction. Some are decimated by destruction of forests, some by draining of swamps, some by building of dams, and some by loss of grasslands. When expanding cities and their suburbs cover the land, there is no chance for wildlife. But as we have seen, the secondary succession from field to forest and from one kind of forest to another may be favorable to a changing group of animals. Often these are kinds sought after by man—deer, for example, which can thrive best along the edges of forests, where meadows meet trees offering abundant shrubs and young trees for browsing.

Plants are continuously on the move, competing with man as well as with each other. Man must manage and control them to gain living room as well as food, and he must care for them to maintain his own environment. Whenever man abandons his cities, plants cover all evidence of his having lived there. They creep among the ruins, prying

Ruins of Machu Picchu, Peru, cleared from the jungle of growth that once covered them. Only continued effort keeps them from being overgrown again. Crustose lichens (the white patches) make a home on the rock surface. Where enough soil has been formed, ferns thrive. (*Henry T. Northen*)

apart the stones and veiling his works of art. They advance rapidly in warm, moist areas, less so in dry regions, but inexorably everywhere. Fortunately, in time plants also cover the ugly scars men leave on the landscape.

It is a source of continuing wonder to us how some kind of plant has always sought out and occupied every available place on earth, how life has persisted in the face of upheavals such as violent eruptions, glacial scourings, floods, fires, tempests and hurricanes, continental shifts, mountain building, and erosion. In the long process of adaptation, forms unable to accommodate to drastic changes have died out and others have evolved to take their place. Even now when a catastrophe eradicates life in some spot, living things move back in as soon as conditions ameliorate. Thus places not now habitable may have once teemed with living things, and may again in some future age.

chapter eighteen

War and Peace Among Plants and Animals

A major strike, failure of a utility system, or the cutting off of supplies by flood or blizzard reminds us with sudden shock how dependent we are on each other. The finely woven web of our civilization is easily broken, but for the most part it is also easily repaired. Another web, far more subtle and far more crucial, exists between the plant and animal life on earth. This web of interdependence functions very well without man, yet we are a part of it. We are also intruders, reaping benefit from it, dipping into it here and there for our life's sustenance, but often upsetting its balance by our ruthless inroads and carelessness. This web is not so easily repaired; some of its threads will remain forever broken, some are slowly healing because we are now cognizant of the threat and are making an effort to undo the damage.

It is interesting to trace what are called "food chains" in the various plant and animal habitats. Creatures that feed directly on plants are the herbivores, those that eat flesh the carnivores, and in each class are some of the smallest and the largest of animals.

Herbivores include mice, rabbits, most insects, deer, elk, moose, elephants, rhinoceroses, hippopotamuses and many others. If all the wild herbivores in a single environment had the same feeding habits, the plants would suffer, and the animals would soon starve. Herbivores that live together usually feed on different species or different parts of a single plant. In Africa, for instance, giraffes graze the high branches, elands munch on tall shrubs and branches at a middle level, impalas still lower ones, and the wart hog feeds on the roots of herbs. Only a few herbivores actually feed in the tree tops—among them squirrels, who concentrate on nuts; monkeys, some of which eat both leaves and fruits; and of course, insects.

Beavers are among the few herbivores that do not accept the environment as it is, but instead modify it by building ponds and lagoons. Their food in summer consists of grasses, sedges, herbs, and various shrubs. In winter they rely on the inner bark of willow, aspen, and other trees and shrubs. They gather the branches during the summer and push them into the mud at the bottom of the pond. During the winter they leave their lodges periodically and drag the branches in.

Carnivores keep the herbivore population in line with the food supply. The big cats are examples of the carnivores; the African lion is the most famous, but is rivaled by the tiger in India, the mountain lion in the United States, and the jaguar of tropical America. Almost as well known are wolves, coyotes, foxes, and weasels; not so familiar is the tiniest and fiercest of all, the shrew. Many birds are carnivores: owls, hawks, and eagles prey on such herbivores as mice and rabbits; and many smaller birds eat insects. Some insects in turn prey on others. Man and other omnivores come into the food chain at various levels.

When the food chain is interrupted, disaster often follows for one or another of its links. Although it seems cruel that one animal should feed on another, where predators are wanting, the population of herbivores may explode, and often greater numbers starve than would fall victim to predators. In the Kaibab Forest of Arizona, the great forested plateau that lies north of the Grand Canyon, the killing of coyotes, wolves, and mountain lions led to an overpopulation of deer. Prior to 1907 these predators kept the deer population between 5,000 and 10,000. Between 1907 and 1923 man upset the balance by putting a bounty on their heads; as they declined the number of deer increased. By 1925 they numbered 100,000, a population that far exceeded the food supply. All forage within reach of the deer was consumed, and during the following two winters 40,000 starved to death. The survivors were left in poor health and the plants were in bad condition as well. The situation has since improved, and the deer are now in about the same numbers as before.

It is not always man that upsets nature's balance. Sometimes the weather will do it. A late freeze in the spring of 1968 caught the nut-bearing trees in bloom. That fall, millions of squirrels in the mountains of North Carolina, Tennessee, Georgia, and Missouri discovered that there were hardly any nuts to gather and store. Feeding on their usual summer fare of green vegetation, they had not become aware of the disaster until it was too late. So strong is their instinct to store up nuts for the winter that, finding none, they began frantic mass migration to any place where they might find them. Like lemmings, they swam rivers and lakes, and many were killed crossing highways as they swarmed into neighboring states. Witnesses to the migration reported that fat and healthy squirrels were on the move, driven not by present hunger but by an instinctive sense of doom.

The pastures of the ocean are formed by the microscopic algae that live in the upper sunlit waters, concentrated near the surface but extending down in lesser numbers to about 600 feet. Microscopic animals graze this pasture, made up largely of diatoms and dinoflagellates, and they are fed upon by forms just visible to the naked eye. These in turn are food for larger ones—tiny fish larvae and little wriggling shrimp. The waters fairly vibrate with this layer of minute living creatures, a congregation called plankton, and through it move larger creatures feeding on it and on each other.

Many of the earth's big mammals—the seals, porpoises, and whales—live in the sea. Of these, the blue whale is the largest animal ever to have lived on earth, three or four times the size of the largest dinosaur, fifty times the size of an elephant. Oddly, the blue whale and others of the baleen type feed on the tiny shrimp or "krill," straining them out of the soup of the plankton. Through the oceans migrate great schools of fish and marine animals, some of them destined for breeding grounds not yet discovered by man, some of them pursuing their favorite fish as they in turn follow their instinctive paths, some led by ocean currents that guide them seasonally to warmer or cooler waters. The migration of gray whales northward off the coast of California, Oregon, and Washington is an annual spectacle. They can be seen from the front porches of cottages and from the rocky shore. But wherever they go, algae must be present to support the food chain.

During the summer, three million Alaskan fur seals move in from the north Pacific to congregate on the Pribilof Islands in the Bering

Sea for their breeding season. There the males eat nothing for several months, while the more sensible females catch fish when they are hungry. The herds spend the winter in the open Pacific, but what they ate there was long a mystery, and was eventually solved in a peculiar way.

Oceanographic ships sounding the depths of the north Pacific often received shadowy signals bounced back from phantom bottoms, levels closer to the surface than the "hard" signals indicated the real bottom to be. Moreover, the phantom bottoms moved up and down, being close to the surface at night and as deep as 900 to 1,700 feet during the day. Timing showed that they began to move down with the first faint light of dawn and up with the coming of twilight.

This moving layer has been found to consist of creatures of the sunless depths, rising as daylight fades to feed at the banquet table of plankton, and dropping back into the dark security of deep water as the sun rises. Some of the creatures are so shy that a bright moon will prevent their rising. Among them are lantern fish, which lure their prey with little phosphorescent lights placed along their bodies. The Alaskan fur seals were discovered to be catching and eating the lantern fish! Thus strange links in the food chains are constantly and often accidentally discovered.

Fragrances in plants are one of their charms for human beings, yet not all fragrances are equally delightful to each of us. The rich odor of gardenias, pleasing to one person, may be "too strong" for another. Ask a number of women what their favorite perfume is and you will find that some like a sweet one, others a spicy kind. An odor that is quite noticeable to some people may be undetectable to others. Such reactions indicate a subtle response of the chemical makeup of individuals. It is not strange, therefore, to find that plants react in various ways to each other's fragrances, or to chemicals we would not consider perfumes at all. Nor is it surprising that such chemicals should be of some use to them in their struggle for existence.

Birds, fish, animals, and insects often establish a home range and keep out intruders. On a larger scale, protecting a territory becomes the activity of a group—honey bees their hive, ants their own anthill, elephants their herd, and human beings their country. Some plants have defensive armor—thorns and nettles, for instance—against animals. Some defend their territory against other plants simply by growing faster and more vigorously than their neighbors. But those that engage in actual warfare use chemicals as weapons.

A eucalyptus grove always impresses visitors with the bareness of its floor; no weeds or grasses cover the ground. A substance produced in its leaves is toxic to most other species, and when the leaves fall the chemical is leached out by rain and prevents other plants from growing. A striking exception is the Monterey pine (*Pinus radiata*), which tolerates the toxic material so well that it can be planted in alternate rows with eucalyptus.

Many species accept black walnut as an ordinary neighbor, living in harmony with it if they happen to be close by. Among the many are peach, plum, pear, black raspberries, mints, violets, goldenrod, ragweed, dandelions, corn, oats, wheat, rye, buckwheat, white clover, and the desirable grasses such as Kentucky bluegrass. On the other hand, substances produced by the roots of black walnut are toxic to some other species, and the poisons remain in the soil to prevent their growth long after a tree is cut down. Among these are apple, azalea, rhododendron, mountain laurel, blackberries, huckleberries, blueberries, tomatoes, potatoes, alfalfa, and, interestingly enough, the undesirable poverty grass and broomsedge.

The black walnut therefore often has growing under it a community of plants quite different from that in the general area; it definitely alienates some species and attracts others. So fond of black walnut are certain species that they form an almost predictable community within reach of its roots. A black walnut growing in a poor pasture, where the ground is sparsely covered with poverty grass and broomsedge, will have a circle of black raspberries around its trunk and a "lawn" of Kentucky bluegrass and clover, sprinkled with violets and mints, while it keeps at bay the undesirable poverty grass and broomsedge. It is true that broomsedge will not grow in shade, but poverty grass grows just as well under maple, beech, and hickory, and these do not attract the better grasses or other plants. Black walnut, therefore, might well be used for improving pasture land. On rocky hillsides in the Appalachian Mountains, where wild blackberries grow freely, the blackberries stop at the root line of black walnut trees, even though they grow under other trees in the area. In an alfalfa field where the land is seeded up to the trunk of black walnut trees, the alfalfa will be replaced with grasses within the circle of their roots.

Chemical warfare is more common in arid regions where sparse rainfall makes both water and nutrients hard to come by. The desert brittlebush and the guayule flourish in solitude because, as with eucalyptus, toxic substances in their leaves prevent the growth of competitors.

Many of the aromatic desert plants, whose fragrance is synonymous for us with the colorful land and the bright sun, use their perfumes to

Chemical warfare. An aerial view of intermixed *Salvia leucophylla* and the sagebrush *Artemisia californica* invading grassland in the Santa Ynez Valley, California. As the clumps move into the grassland the terpenes they produce kill the grasses ahead of them, keeping a bare area around their borders. (*Cornelius H. Muller*)

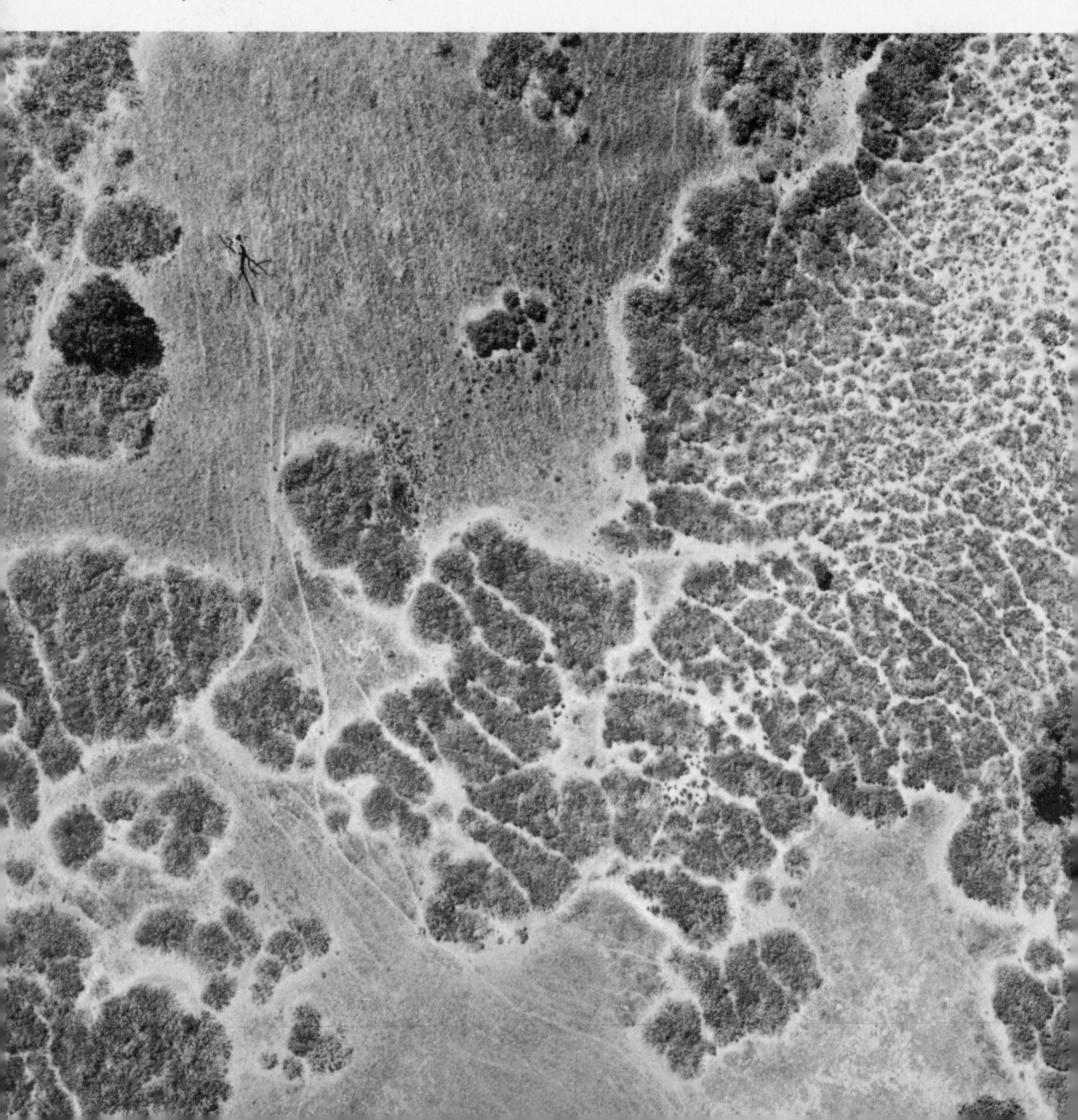

hold other plants at bay. In southern California the sagebrush, *Artemisia californica,* and salvia, *Salvia leucophylla,* live together in thickets surrounded by a 6- to 8-foot zone which they keep denuded of grasses and weeds. Their own seedlings do grow in this bare zone, and the thickets thus advance on all fronts, killing out the grasses ahead of them. Their fragrance consists of volatile chemicals called terpenes, which are taken up by the soil and kill seedlings of grasses and weeds. Ironically, the terpenes in heavy concentration are apparently toxic to their own seedlings, because the young plants seem to be able to grow only outside the thicket, where they receive the terpenes in diluted form. Even the oldest plants may be affected, for in long-established thickets, the ones in the center slowly die out. As the older plants die in the center and young ones grow in the flanks, the thicket moves into new territory, invading and pushing back the grasses. A time-lapse aerial movie would show the thickets creeping from place to place, nosing into the grassland, sometimes breaking up into new colonies.

Detail showing denuded area between the salvia on the left and grasses on the right. At *A* all the grasses have been killed. At *B* the grasses are stunted. At *C* the grasses are still unaffected. (*Cornelius H. Muller*)

It has been conjectured that the terpenes of these species are products that the plants do not use in their own physiology, and that may be harmful to them if retained or if fed back to them in quantity. By excreting the terpenes, the plants get rid of them, and at the same time control surrounding plants so as to maintain their own territory. It is a strange story of two kinds of plants living together, sagebrush and salvia, tolerating the materials they both have to get rid of, and at the same time putting them to use to make survival easier for each other.

Other desert species have been found to practice herbicide of a similar nature; for example, *Artemisia tridentata,* the sagebrush of the central plains, two other species of salvia, *S. apiana* and *S. melifera, Heteromeles arbutifolia, Prunus ilicifolia* and *P. lyoni,* and *Umbellularia californica.*

Evergreen trees excrete terpenes in amounts so great that a blue haze may cover the forest. They give forests their refreshing odor, the fragrance we enjoy bringing into our homes at Christmas time. There are hundreds of terpenes, all different chemically. Their role in nature has not been determined. There may be among them other herbicides; some may function as insecticides. Investigation into their possibilities has only just begun.

There are estimated to be about three million species of insects on earth, in numbers that stagger the imagination. From man's point of view, a few thousand are major pests, kinds that spread disease among animals and human beings and harm the plants we use and enjoy. They take a tremendous toll of farm products each year, in an age when food is desperately needed.

Some trees maintain armies to defend them from their enemies. Species of *Acacia* have hollow spines that make convenient homes for fighting ants, and their leaflets secrete small egg-shaped globules rich in oil and proteins on which the ants feed. Since they have both shelter and food, the ants need never leave the trees. They viciously attack any intruders that threaten their boarding house, even man. Among the enemies of the acacias are the leaf-cutting ants, which would quickly strip it of leaves if they were not driven off by the defending army. *Cecropia* is another tree that maintains ant armies, housing them in hollow stems. If the fighting ants are killed, it isn't long before the leaf-cutting tribe sets to work.

Leaf-cutting ants were farmers long before man. Their sole food consists of pure cultures of fungi which they grow in subterranean gardens on bits of leaves. So busy are they cutting leaves that they wear little highways through the forests, along which a two-way procession travels all day long, one line bearing aloft pieces of leaves like sails, larger than the workers themselves, the other returning for more. In any one underground garden only one kind of fungus is allowed to grow, in spite of the fact that spores of many kinds cannnot help but enter. How the ants control the purity of the culture is not known, although it is conjectured that their excreta or saliva contains antibiotic substances that inhibit all but the chosen kind.

Ant farmers and their fungus garden. The ants maintain a pure culture of the fungus, which provides them with food. (*Neal A. Weber*)

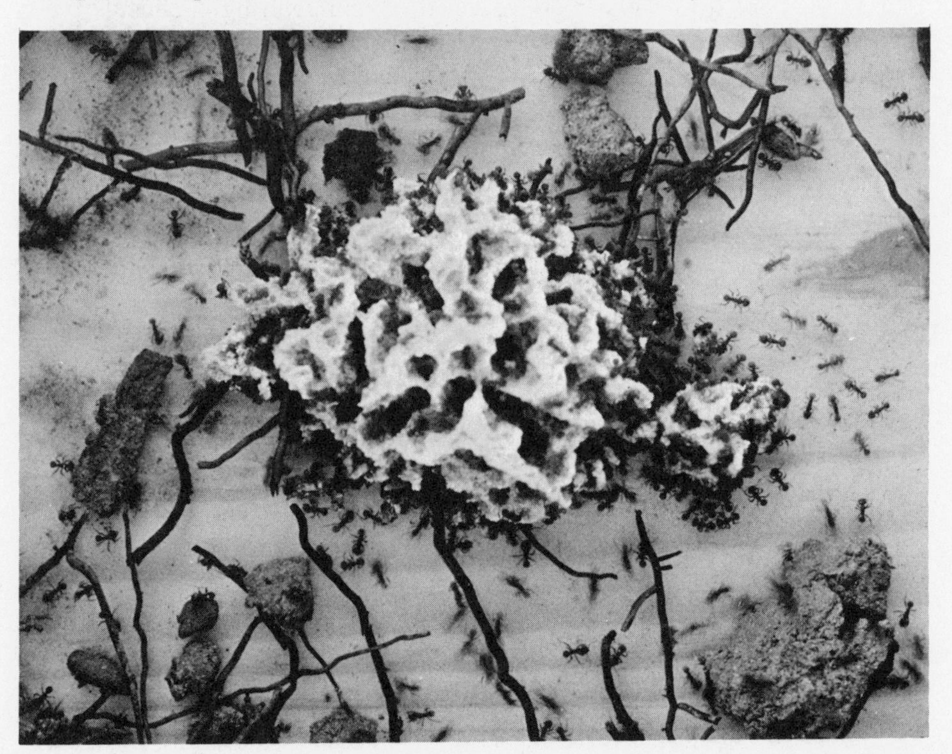

The ants also manage their gardens so that the fungi behave differently from their normal habit. The usual mycelium, the mass of threadlike filaments that serves as "roots," is produced, but it is prevented from forming fruiting bodies, or mushrooms. Instead, the ants control it so that the tips of the filaments produce little white globules called bromatia, which are produced only under their cultivation. It is the bromatia that the ants feed upon. New gardens are started by young queens, who take a bit of the fungus with them when they leave the nest. They nourish it on their own excreta while they lay their first eggs, and feed it to the developing workers until the latter are grown and numerous enough to begin farming.

No one kind of plant has been able to fend off all its enemies, but it usually reduces the number of kinds that will attack it. The fact that most insects prefer some plants over others suggests that the ones they do not choose are unattractive to them in some way. Indeed, many plants defend themselves from insects by manufacturing chemicals that are either distasteful or poisonous. One of the earliest insecticides used by man is made naturally by the plant *Pyrethrum;* the chemical pyrethrin is still used today. Plants that manufacture strong alkaloids are particularly well protected from insect enemies, among them tobacco (nicotine), cinchona (quinine), coffee (caffeine), strychnos (strychnine), coca (cocaine), hemp (marijuana), opium poppy (morphine), and peyote (mescaline). It will be interesting to learn in the future just what effects these substances have upon the few insects that do feed on them, and whether they may cause addiction or psychological derangement!

Man's battle against insects has backfired in many ways. Most of the chemicals used to control them are not selective and will kill off useful ones and affect other forms of life as well. The warnings given by the late Rachel Carson in her book *Silent Spring* have come woefully true. Birds that feed on insects and fruits that have been sprayed have been decimated; lakes and rivers have been polluted, fish killed, and waters made poisonous to human beings; and death and illness have been caused among people and animals. Insects multiply so rapidly that mutations are frequent, some of which have led to resistance to insecticides. Also, some insects apparently have a way of detoxifying poisons used against them. The search for controls now leads in two directions—of finding chemicals that will affect specific insects and not harm other forms of life, and of finding kinds to which they cannot become resistant.

Insects produce a series of hormones that control their development. One controls the insect's immature stages and is necessary in order for the insect to develop properly. When the larval stages are completed, production of this so-called "juvenile hormone" ceases and others controlling the adult stages begin to function. When research workers discovered the juvenile hormone and applied it in continuous doses to larval forms, they found that the larvae could not change into adults and died without maturing. The hormone was first extracted from cecropia moths and their relatives the cynthia moths, and oddly, these are the only insects from which it has been possible to obtain it. Although the hormone is extremely potent and a tiny amount will fatally throw off the internal timetables of millions of insects in the larval stage, not enough can be obtained to control the huge pest population. Also, since it is not selective, it kills useful insects as well.

A most fortuitous event opens up the possibility that plants may aid not only in their own salvation but in ours. An entomologist from Europe brought with him specimens of an insect, *Pyrrhocoris apterus*, on which he had been experimenting. Attempts to rear them in this country failed; they either died in early larval stages or developed into abnormal larvae that could not shed their skins and eventually died. None matured. Apparently they were getting juvenile hormone from some source in the laboratory, and this source proved to be the paper toweling in the cages! Investigation showed that many kinds of paper, including newspaper, caused their death, and further search narrowed the mystery down to paper that contained pulp from the balsam fir. Analysis of chemicals made by that tree revealed that balsam fir produced, among its many terpenes, one that differed only slightly in molecular structure from the known juvenile hormone (and therefore called a juvenile hormone analogue). Amazingly, it was effective only against this particular insect. Species of *Pyrrhocoris* are not now enemies of the balsam fir, but the researchers speculate that some members of this family of insects may have been its enemies in the dim past and may have been destroyed by ancestors of our present balsam fir. Even though the tree does not need the protection today, it still makes the juvenile hormone analogue.

The family Pyrrhocoidae contains some members that attack cotton plants. It is not yet known whether the substance produced by the balsam fir will kill these, too, but the clue is being followed up.

Other plants have now been found to produce chemicals similar to juvenile hormones, among them some weeds. Those produced by different plants are just slightly different from each other chemically, but they result in death of the insects without their maturation. The bracken fern makes two such chemicals, which act in slightly different ways, but with the same ultimate result. As they are analyzed in the laboratory, it may be possible to make synthetic ones that will affect specific insects. Thus plants that produce their own insecticides are not only helping themselves but are giving us weapons to control harmful insects without killing beneficial ones.

While some plants seem able to fend off most of their enemies, others actually invite disaster. The mulberry, for instance, attracts silkworm moths and feeds their larvae. The red oak performs a similar service to the polyphemus moth which feeds upon it. Without a certain volatile substance produced by the tree, the moth cannot mate. The female, which is unable to fly, relies on sending out a perfume to lure the males to her, but she can send out her attractant only after her antennae have received the stimulating chemical from the red oak. Even though male and female moths are kept together in the laboratory, no mating takes place until leaves of the red oak are brought in.

The chemical was found to be trans-2 hexanol. It has also been found in maple, birch, beech, and elm, but strangely, none of these provokes mating in the moth. Apparently the trees make a second chemical that masks their trans-2 hexanol. Also, all except the elm cancel the effect of the red oak, although why the elm fails to do so is not known. Curiously, a number of other volatile substances that were tried in the laboratory were also found capable of blocking the red oak factor, among them Chanel #5! It hardly seems possible, however, that this expensive scent will ever be used to protect the red oak from the injury of the moths it continues to aid and abet.

Sex attractants are being put to work by man to deceive the males of other insect species. Isolating the particular chemicals from the various insects is not as easy as it might seem, but it has been accomplished for the gypsy moth, the cabbage looper, the pink bollworm moth, the army worm, and the American cockroach. Although the sex attractants are similar chemically, each is just enough different from the others to work only on the species that produces it. The gypsy moth lure was the first to be synthesized. When it is used to lure the males into traps, the females are denied their mates and produce no offspring. A tiny bit of the attractant is put in the trap, which holds either a sticky substance that entangles the males that fly into it, or a poison that kills them. The former is the safer method because the trap can then harm no other living thing. The traps are dropped from airplanes into forests and groves. Similar methods have already proved successful for the cabbage looper and the pink bollworm moth. Housewives will be overjoyed when they can buy traps to eradicate cockroaches. The beauty of the system is that the insects can hardly become immune to their own hormones or sex attractants.

Other attractants that have nothing to do with sex have been found to lure male insects—and only the males—although we don't know why. Chemists find them by testing an assortment of fragrances at random until they find one that works on some destructive insect. They now have scents that will entice the Japanese beetle, the European chafer, and several kinds of fruit flies to their doom.

The story of finding one for the Mediterranean fruit fly that recently threatened the Florida citrus crop is particularly intriguing. By trial and error, it was found that oil from the seeds of a species of *Angelica,* a very pretty wild plant, attracted the male fly—but only a small amount of the oil could be produced. In the emergency that faced the Florida citrus growers, anyone who could help went to work. The perfume industry made up a compound that smelled exactly like angelica seed oil, but alas, it elicited no response from the fly. At the same time chemists made a synthetic attractant and modified its structure until they had five or six compounds. Not one smelled like angelica oil, and all smelled differently to the human nose, yet they *all* appealed to the fly. So it was not the *odor* of angelica seed oil that lured the Mediterranean fruit fly, but something the insect found entrancing that could not be detected by human beings—another proof that a fragrance can have different effects on different creatures.

A "birth control" approach is also being employed. Male insects are reared and sterilized by irradiation, then released to compete with wild ones. For every mating between a sterile male and a normal female, some thousands of eggs come to naught. Thus several selective methods are now available, none of which can harm anything but the pest at which it is directed.

Insects are pretty tricky. Just as they have been able to develop resistance to insecticides, so have some developed a tolerance, even a liking, for some of the distasteful and occasionally poisonous chemicals plants manufacture. In doing so, they may also make themselves unpalatable to would-be predators.

The monarch butterfly prefers to lay its eggs and have its larvae feed on members of the milkweed family, which contain glycosides—poisonous substances some of which are used to treat heart disease in man. Birds will not eat the monarchs, since they cause immediate vomiting. Research workers offered them to birds raised in captivity and denied any contact with monarchs. The birds tasted them—once—and firmly rejected them thereafter. When offered monarchs raised on cabbage instead of milkweed, the same birds still refused them on sight. A second group of birds that had not yet experienced monarchs happily ate those raised on cabbage. Clearly the flavor of the adult butterfly depends on its diet as a larva, and the butterflies, made distasteful by the chemicals of the milkweed, are protected from their enemies.

The viceroy butterfly takes advantage of the monarch's offensiveness and escapes being eaten by mimicking the monarch in appearance. Part of the experiment just described was to offer viceroys to the two groups of birds; those that had once tasted the offensive monarchs refused the viceroys, whereas those that had not, showed no hesitancy in eating them. Other butterflies mimic other unpalatable species to escape being eaten.

There are some insects that protect themselves by imitating plant parts such as leaves, even dead ones, thorns, or twigs. The praying mantis and the walking stick are most familiar. They are practically invisible on the plant they mimic as they wait for unsuspecting prey. One mantis mimics the pink Malaysian orchid *Spathoglottis plicata* not only by being pink but also by having petallike ruffles on its legs.

Insects may cause little harm in their native countries, where their population is checked by natural enemies, but they can wreak havoc when introduced elsewhere. The inadvertent introduction of the gypsy moth into forests in eastern states has caused much devastation. It was brought in from Europe in 1868 by a naturalist who accidentally let some escape. It spread unchecked and its larvae ravaged hundreds of thousands of trees. Help was sought from European scientists, and after studying the moth and its natural predators, they found that the Calosoma beetle was one of its major enemies, eating both the larvae and the adults. The beetle was introduced into the endangered forests, and

This treehopper insect gains protection by mimicking a thorn. (*Gerald Lang*)

although it has not eradicated the moths, it has been helpful in reducing their population.

Undesirable plants sometimes make their way to our shores. The Klamath weed or goat weed—native to Europe, where it is known as St. John's wort—was first reported in Pennsylvania in 1793. By 1940 it had ruined 250,000 acres of grazing land in California. It also invaded Australia and was causing the same problem there. Its damage is two-fold: it drives out the grasses by multiplying during the dry season when the grasses are not able to compete with it, propagating itself both by runners and by fine, wind-borne seed; and it is injurious to cattle, causing a sensitization of the skin of animals with light-colored skin or white spots. The animals develop painful sores and become irritable and hard to control.

The Australians imported insects from France—a leaf-eating beetle and a root borer known to feed on this weed—after checking them thoroughly to make sure they would not damage other plants. The United States obtained them from Australia after making further checks on our own native species. Fortunately, these insects were found to feed on the Klamath weed only, and they have now nearly killed it off. They work neatly as a team: the beetle prefers plants in the sun and the root borer those in the shade, and the borer attacks the Klamath roots during the dry season when the beetles are hibernating. As the weed declines, the population of these two insects also declines, and if it is ever entirely eradicated, the insects will die out, too.

Over 200 of our most noxious weeds have come in from other countries without being accompanied by their enemies. Controlling them by bringing in predators that feed only on specific plants looks very promising.

The United States Department of Agriculture quarantines all plants brought into this country and inspects them for diseases and pests. Diseased plants are discarded; those harboring pests are fumigated. Certain kinds are forbidden entry because they may bear organisms that would be harmful to an important crop—for instance pineapples to Hawaii. A small mite (called a false spider mite) which came in unnoticed from Central America caused much damage to orchid plants in this country before it was detected and controlled. Animals and birds are also inspected and sometimes quarantined before being allowed entry—for instance, cattle because of hoof and mouth disease, and parrots for parrot fever. A traveler who smuggles in a foreign plant or animal may do himself or his country a great disfavor, for although he may never realize it himself, something he introduces may cause damage eventually.

epilogue

Guarding Our Environment

FAR too many of us have thought of nature as separate from ourselves—a forest a thousand miles away, a remote lake, a kind of bird we never saw. We have gone to our jobs, enlarged our cities, ploughed more land, cut trees, used the nearest streams to haul away our wastes and those of our factories. Suddenly we discovered that nature is not apart from us, but as close as the air we breathe. We know now that this earth is a sanctuary for all living things, not just for man; that when we make it unfit for plants and animals we also make it unfit for ourselves. Those of us who were willing to save nature for herself alone now realize that we *must* save her in order to save ourselves.

We'd like to have the skies blue again over our cities, not filled with smoke and poisonous chemicals. We'd like to catch trout in pure running streams, and collect sea shells on clean white beaches. We'd like to see bountiful grazing lands where there is now erosion and bare ground, and waving fields of grain where there are now farms without topsoil. We used to have flocks of bluebirds in the spring—we'd welcome them back. There are still places in the country where we can have all this, but we have to go far to reach them. And as we grow in numbers, those places shrink in proportion.

The problem of air pollution must be solved or man himself may become extinct. There are millions of people who have never had a breath of fresh air since their birth. We know that air is no longer free, and never will be, but we cannot benefit from money spent in other directions if we do not have clean air. Air that is clean at this moment will inevitably, and not too long from now, be filled with pollutants from other areas. If the sources of pollution could be controlled so that no more noxious gases were poured into the atmosphere, the poisons present would eventually be washed to earth and be degraded by biological activity.

The air is fouled by the combustion of fuels; the evaporation of solvents; the gases, dust, and smoke released from industry and from the burning of trash; the aerial dispersal of insecticides, and by many other sources. A few of the resultant contaminants are carbon monoxide, sulfur dioxide, ethylene, aldehydes, peroxyacyl nitrates, nitrous oxides, and fluorides. These are bad enough, but when the sun acts upon them, additional poisons are formed by chemical change. They are especially concentrated by air inversion when visible smog forms, but the air is not free from them even when it appears to be clear. It is not only large cities that are subject; small towns nestled in hollows can have equally serious pollution from just one or two small factories or pulp mills.

The polluted air irritates eyes, nose, and lungs; worse, it is a contributing factor to bronchitis and emphysema. The number of cases of human illness rises in direct proportion to the increase in pollution. In addition, pollutants reduce visibility, create driving hazards, corrode metals and tires, destroy masonry, and dirty clothing, homes, and furnishings. The cost is tremendous. Not measurable in dollars is the aesthetic loss, the shutting out of the sky and obliteration of the view.

Plants are injured even more rapidly than animals. Crop losses amount to at least $12 million a year in California and $18 million in the area between Washington, D. C. and Boston. Farmers have had to give up certain crops or cease operating near urban centers. Damage to plants in parks and gardens can scarcely be estimated. Growers of cut flowers have had to move their greenhouses out of many cities. Trees along highways are becoming sick and even dying; the white pine is especially sensitive. Orchid growers in certain areas are now suffering considerable damage to their flowers. In fact, the Cattleya orchid is

Tulips injured by sulfur dioxide, a universal pollutant over metropolitan areas. Fluorides in the air cause the same type of injury. (*Robert H. Daines*)

blasted by such low concentrations of ethylene gas that it is used as an indicator of this poison's presence in urban areas.

Symptoms on other plants have been found due to pollutants. Ozone, for example, causes white flecks on tobacco leaves, aldehydes cause white bands on petunias, sulfur dioxide scorches the leaves of tulips, and ethylene causes poor growth and flower drop on carnations and snapdragons. Crops and forage plants are similarly injured—alfalfa, cotton, spinach, beans, lettuce, and tomatoes. Some of these, too, have been used as indicator plants to warn of the presence of specific pollutants.

Attempts are being made to cut down air pollution. Cities are curbing the incinerator burning of trash and are setting standards for industry. Inefficient furnaces and boilers are being replaced with modern ones that burn fuel more completely, right down to the carbon residue. High stacks, forced drafts, electrical precipitators, and scrubbing towers reduce the amount of material discharged into the air.

Often these devices pay their way because valuable by-products previously emitted may be recovered. Los Angeles, plagued with smog that affects 80 percent of the population 200 or more days a year, has pioneered in pollution abatement, yet it is just keeping pace. Although its air is just as bad as it was ten years ago, if there had been no control it would be far worse. We've already spent a billion dollars to combat air pollution, but as we lessen pollution from present sources, new ones come into being. Increased population, more cars, and new industries offset the progress we have made. Elimination of pollutants from the sources that have been controlled is obviously not enough, and the reason why makes us all culprits.

By any standard, the automobile is the major polluter of the atmosphere. The carbon monoxide and nitrogen oxides given off by cars measure ten times greater than poisons from all other sources. In Los Angeles, motor vehicles spew 4,439,490 tons into the air each year, and in Denver, 600,000 tons. Hydrocarbons given off by tires equal 100 tons a day in Los Angeles. The number of tons of car-based pollutants so closely approaches the figures of population that it is apparent that each of us who drives a car or truck is responsible for about a ton of poisons a year. Newly developed engine and exhaust modifications cut down the unburned petroleum from cars, but do not reduce the nitrogen oxide.

Most pollutants in the air are invisible, but if they were as visible as garbage we would recognize the disposal problem. The difficulty is that no human efforts can remove this garbage from the air once it has been placed there. A big step in alleviating air pollution would be to reduce the use of cars or find new types of engines and power sources, or to install alternative means of public transportation. The old trolley car that we so hastily did away with was a greater boon than we know.

Combustion of fuels in homes and industry adds great quantities of carbon dioxide to the atmosphere. Since green plants use carbon dioxide in photosynthesis, it might be assumed that additional amounts would be beneficial, but in fact, the plants we have left now cannot use it all. It accumulates, and if the current rate continues, it is estimated that by the year 2000 we will have 25 percent more carbon dioxide in the atmosphere. The higher concentration may affect the world's climate, producing a warming trend that will melt glaciers and raise the level of the ocean to the point of inundating our coastal cities.

It is a strange situation indeed: as cities increase in size, the plant cover is reduced. As the population grows, more oxygen is needed, but we are paradoxically cutting down the very source of our oxygen supply. The move to create parks within cities has been promoted to give relief from concrete and glass and provide spots of beauty and places where apartment-raised children can play. The human spirit does need the lift that plants and space can provide, but creating more parks can also help forestall literal suffocation in urban areas.

The greatest threat to the world's oxygen supply, however, is damage to the oceans. Even such small concentrations of DDT as 0.6 to 6 parts per billion of water reduce the growth and photosynthesis of marine algae, which manufacture more than half of the world's oxygen.

While we are intent on cleaning our atmosphere so that we can breathe, we have condoned (partly through ignorance and apathy) the manufacture and storing of poisons that could eliminate mankind completely. We were alerted to the magnitude of the danger by the now well-known killing of 6,000 sheep during test-spraying by the Army on the Dugway Proving Ground in 1968. Our feeling of shock was intensified when it was learned that winds could drift the gas from continued testing to two towns and a public highway not far away.

It has since come to light that the Army has stored enough poison gas to "kill the world's population ten times over." People in cities where the depots are located (in Denver, next to a major airport) object to being such ready targets. The accidental bursting of *one* canister is said to be capable of killing a million people. The Army excuses itself by saying that the possibility of accidents is "remote," but the newspapers reveal how often wrecks occur on railways and highways—and not only frequently, but disastrously around airports. Thus both storing the gas or moving it from place to place is fraught with frightening possibilities. The proposed use of such poisons in warfare is appalling enough, but we might not even have to wait for a war!

Water and life have been inseparable since life began in the ancient seas. When organisms invaded the land, they came to rely upon precipitation for their water. The sun desalts the ocean water as it evaporates it. As air carrying the moisture rises over the mountains, or is lifted in frontal disturbances, or rises in unstable air masses, it is cooled, forms clouds, and drops the moisture back on the land. Some of the water enters streams directly, but most enters the soil. In humid regions some may descend to the water table and seep into streams which carry it back to the ocean, but most of it returns to the air through evaporation from the upper soil layer or by way of plants. Of the water absorbed by roots and moved through the plant, only about one percent is used in photosynthesis; 99 percent is evaporated from the leaves in transpiration. Some of this may later fall locally as rain or snow. However, much of it remains in the air as it circulates back over the oceans. There the air is charged with more vapor and moves again over the land. The water that falls as rain or snow is pure unless it gathers pollutants added to the air by man.

As our industrial society expands, it becomes increasingly difficult to supply water for everyone. Man often settles in deserts and demands that water be brought to him, or he builds on flood plains and demands that water be kept away. Huge aqueducts carry water to southern California and Arizona, and people in these states now look thirstily and enviously at the Pacific Northwest and British Columbia for more. Excess water that floods the river valleys passes by, doing more harm than good, and escapes without being put to use. We use water lavishly and wastefully—about 200 gallons per day per person. In addition, enormous quantities are used for irrigation and industry. Ten gallons are used to make one gallon of beer; 18 barrels to refine one barrel of oil; 250 tons to make one ton of paper. The total use of water today in the United States amounts to 320 billion gallons a day. By the year 2000 we will need 900 billion gallons a day, which exceeds the 700 billion gallons now available, and we will have to go to the oceans for the difference. The price of desalted water should by then be reasonable. The mountains of salt that would be obtained as a by-product could be the source of valuable minerals—sodium, chlorine, boron, magnesium, potassium, bromine, silver, and gold. Even so, much of the brine would have to be returned to the ocean, with care not to injure marine life.

Our lakes and streams have become dumps for domestic and industrial wastes, with consequent poisoning of the water for human use, harm to aquatic life, and loss of scenic beauty and recreational value. Waterways near some urban centers are loaded with domestic sewage, wastes from pulp and paper mills and oil refineries, wool washings, offal, and acid, lime, oil, and grease from steel plants. Most are also contaminated with insecticides, most dangerous of which are the chlorinated hydrocarbons such as DDT, aldrin, dieldrin, heptachlor, Lindane, and endrin—kinds that are persistent and almost indestructible. Some food-processing plants wash the residues of poisonous sprays from fruits and vegetables and pour the poison-laden water back into the rivers. Rains wash pesticides into rivers from sprayed fields. Massive fish kills have been brought about, for fish are especially sensitive to insecticides.

The poisons go on to the ocean. That dilution in the ocean reduces the hazards is a myth, for marine organisms are far more sensitive than land creatures. Also a myth is the idea that substances do not travel far in the oceans, for currents flow from east to west and north to south. Migrating fish and mammals carry the DDT they obtain from organisms they eat, and creatures that eat them in turn acquire the DDT they contain. DDT has been found in the bodies of penguins in Antarctica. Damage to the plankton, the algae and crustaceans that are the basic food for all other marine life, spells potential disaster to life in the seas. There follows naturally great economic loss from contamination of seafoods, as well as threat to human health.

We cannot tolerate pollution; we must have pure water. In 1965 Congress passed the Water Quality Act, which set up Federal standards and guidelines for the return of water to streams and lakes, and in 1966 the Water Restoration Act, providing money to aid communities in meeting these standards. Some sources of pollution were overlooked at that time—specifically, acid pollution from mines, accidental oil spills, oil leaks from offshore wells, and discharge of pollutants from commercial and pleasure ships. Legislation to control these is under way. A start has been made, but many years will be required to correct past mistakes and keep up with current sources of pollution. For sewage-treating facilities alone, at least two billion dollars a year will be needed for many years. Enforcement leaves something yet to be desired. About 1,550 communities in the United States, with a total population of 9.5 million people, dump 260 million gallons of raw sewage daily into lakes, bays, and rivers. Even in cities that have sewage plants, the facilities are inadequate in 5,000 cities out of 12,000. There are still industrial contaminators who slip wastes into the nearest stream when no one is looking. If the persons responsible were to turn on their faucets one day and have these wastes return to them as drinking water, they would cry for help.

We have the technology to free water from organic pollutants from both industrial and domestic sources. We need only apply this knowledge. Bacteriological processes can be used to change harmful substances into innocuous ones. By selecting appropriate species of bacteria, and by providing ample oxygen, most organic pollutants can be degraded, especially if the water is impounded long enough before it is returned to the waterways.

The effluent from sewage disposal plants has high concentrations of nitrates and phosphates, both of which are essentially fertilizers. As we saw in an earlier chapter, when the effluent is released into lakes or streams, the ecology of the water is changed by their enriching influence. Algae respond by making tremendous growth, and they become objectionable by their very quantity. The natural death process of the algae gives abundant food for aerobic bacteria, and their population in turn explodes, using up the available oxygen. Anaerobic bacteria then increase and bring about putrefaction with resulting foul odors and the death of aquatic plants and animals, including game fish. This is what has happened to Lake Erie and is now happening to other bodies of water. Experiments are being carried on in which the effluent from sewage plants, after being purified of harmful substances, is sprayed on forests and crops. The plants make good use of the fertilizer, and money can thus be saved. The surplus water seeps to the water table, where it can be recovered by wells.

As long ago as the 1930's, some of our beaches were contaminated by oil, a slow accumulation drifting in from ships and small boats. On some coastlines the sand grains were covered with a film and would become packed on feet and shoes; on others black globs were present here and there. The situation was unpleasant enough to cause abandonment of the beaches, and though no one liked it, nothing was done about it. No one foresaw the dreadful calamities to come, and which are already being repeated. One of the worst was the sinking of the tanker ship *Torrey Canyon* in 1967; 850,000 barrels of crude oil spilled into the Atlantic off the shores of England and France. As the oil swept relentlessly toward the beaches of Cornwall and Brittany, detergents were dumped upon it in the hope of dispersing it. Then men had to combat both the suds *and* the oil as they were washed onto the beaches. Biologists assaying the damage to marine life determined that the detergents had caused as much, if not more, harm than the oil, and that they hadn't done much to disperse the oil in the first place.

Biologically purified effluent from sewage contains minerals that make a rich fertilizer. Rather than being poured into streams or lakes where it brings about eutrophication, it can be used advantageously on land plants. Here is a winter scene where it is being sprayed experimentally on forest land. (*USDA Photo*)

Another calamity was the runaway oil well in the Santa Barbara channel early in 1969. Drilling in a fault zone (against the advice of experts and the protests of residents) had opened up a crack from which flowed 21,000 gallons a day for many days before it was brought under control. This time the public was aroused not only by the damage to the beaches but by the pitiful helplessness of thousands of grebes and other shore and marine birds that were caught and killed by the oil.

Further damage to marine life will undoubtedly show up; it was not long before migrating gray whales, which travel just offshore, were found with their plankton-straining baleen gummed up with oil, making it impossible for them to feed. Methods of control include legislation that would place the responsibility for cleaning up the mess directly upon the companies whose ships or wells caused the spills. Laws should demand more thorough exploration and study of the structure of the ocean bottom where wells might be drilled, and insistence that the information be carefully used; and research into ways to remove the oil from future accidents and assurance that chemicals used will not injure marine life.

Pollution by heated water—thermal pollution—is just now being recognized as the source of a great deal of damage to lakes, rivers, and bays. The heated water kills some aquatic life directly and other forms by upsetting the environment so that they cannot survive. For instance, some fish die when the temperature of the water goes above a certain range, others die of starvation when organisms on which they feed are killed. The water's capacity to hold dissolved oxygen declines as the temperature rises, and this brings about oxygen starvation for many forms of life. Lack of oxygen also reduces the numbers of aerobic bacteria, which are needed to dispose of waste materials, and anaerobic bacteria increase. The warmed water accelerates the action of chemical pollutants, intensifying their action.

It is now thought that the death of hundreds of millions of alewives in Lake Michigan in 1967—the worst fish kill in the history of the United States up to that time—was caused by thermal pollution and oxygen starvation. The annual run of salmon and steelheads in the Columbia River and its tributaries is threatened and may be destroyed by the Hanford nuclear reactor's warming of the river. The fish suffer upsets in their metabolism, increased susceptibility to disease, blocking of the spawning urge, and premature ripening which brings about failure to spawn. Chesapeake Bay, San Francisco Bay, and other places are in an equally dangerous situation.

Electric power plants discharge 50 trillion gallons of heated water each year. Those using nuclear power are particularly bad because *all* the heat they generate is released into the water. With the rapid increase of need for electric power, the situation becomes worse every year. Ways must be found to cool the water before it enters our waterways, and these techniques must be enforced. With ingenuity, perhaps the hot water can be put to some constructive uses during which it would be made to work while being cooled. Otherwise devices such as cooling towers must be employed.

The problem of radioactive wastes is the most difficult to solve, for the radioactive material accumulates relentlessly in both plants and animals. Their effects will reach far into the future.

In the state of Washington, the water of the Columbia River that is used to cool the Hanford nuclear reactor becomes contaminated with radioactive zinc. From the reactor the water flows back into the river. Downstream, farmers irrigate their fields, and the radioactive zinc accumulates in the hay and other crops. Cattle eat the hay and the radioactive zinc enters their bodies and their milk. Vegetables and fruits raised on the land absorb it. When people eat these things or drink the milk, radioactive zinc becomes incorporated in their bodies. The contaminated water flows on to the ocean, where the radioactive material becomes diluted, to be sure, but the dangerous isotope is accumulated in plants and animals of the sea. Radioactive zinc has already been found in oysters and clams as far south as Coos Bay, 200 miles from the river's mouth. Ocean currents will eventually distribute it further. The bacteria do the best they can with normal pollutants but now man has imposed an insurmountable—and indigestible—problem upon them.

It seems to the casual observer that every third vehicle on the highways is pulling a trailer or a boat or is loaded with camping equipment. No one can tell how many cars carry a picnic basket or fishing rods, or just a suitcase or two. People are out to see the country, and we can be thankful that there is still a lot to see. Cold weather doesn't stop them—the ski areas attract their fans, and the warm climates theirs. People may earn their living in the crowded places, but they seek the beauty spots of the land for recreation. There is an urge to spend some time with nature, to expand under her relaxing influence, to rest city-fatigued eyes with sights not made by man. No country vacation is complete for a child without a glimpse of a deer or a hawk or a chipmunk, and getting close enough to photograph them is just as thrilling to adults.

Now as never before we are determined to preserve the enchanting beauty of our land, and to make it available to our people. In order to do so we must manage to keep the still unspoiled places intact while expanding our population; we must make room for the wild creatures while finding room for ourselves. But we must not merely save the scenic spots for those who can reach them; we must create beauty where there is now ugliness by supplanting the slums and the drab and rundown parts of our cities with better dwellings and more attractive surroundings. Nor can we survive the tensions of a year by having a mere two-week glimpse of wild nature; we need parks around us. Neighborhood parks actually receive more recreational use per hour and per day than the national and state parks, and there aren't yet enough of them.

The term recreation does not mean just play, it means what it says, re-creation. It means to create anew, and if we may add a thought, to bring human beings closer to the state when the race was young and all ahead was fresh and new.

The most spectacular and most scenic places in the country are set aside in 34 national parks. Few of us realize the vast acreages they embrace. Yellowstone is the largest, with two and a half million acres; Mt. McKinley is second with almost two million; and these are followed in order by Everglades, Glacier, Olympic, and Yosemite. North Cascades and Redwoods National Parks are the newest, and Canyonlands is still being developed. A great deal of country seldom seen by visitors surrounds the focal points in each. The 88 national monuments are smaller, but each has a feature of special interest centered in a protected area. Just recently, seven national seashores have been withdrawn from commercialization to be saved in their natural condition.

The story of Redwoods National Park is one of sheer devotion on the part of many people, who formed the Save-the-Redwoods League as long ago as 1918, in order to preserve the area along the north California coast where the largest and oldest of all the redwoods are located. For fifty years they worked to spread word of the danger that threatened the majestic trees and to gather funds to buy the land so that they would be safe from the woodman's axe. Thousands of people contributed a total of $13.5 million. Many individuals bought and preserved small private plots, often dedicated to the memory of some loved one. Their efforts were rewarded in 1968, when Congress passed a bill establishing the 58,000-acre park and authorized $98 million to reimburse owners for land it encompassed.

Within the national parks and monuments we can see volcanoes—some extinct and some likely to be active at any time—great canyons, magnificent mountains, glaciers and snow fields, rainforests, lakes, rivers and waterfalls, far-spreading forests, vast deserts, fantastically eroded landscapes, the bones of prehistoric animals, and petrified trees. In those set aside to preserve the works of man, we can sit on the terraces of ancient dwellings and picture the lives of those long-ago people who knew not one modern convenience. We can see in our minds the men returning from the hunt and the women weaving baskets and decorating pottery in such lovely designs that they are now being copied in modern fabrics, trace out their gardens, listen to the water flowing in the streams from which they drank (but *we* can't drink of them now because they are polluted), and breathe the still clean air they breathed. We cannot help but contrast our lives, in which we buy everything we need, with theirs, in which ingenuity and an intimate knowledge of the land and its plants and animals gave them their living. We feel the tug of the common thread that runs through all humanity and it draws us closer to nature, on which our lives depend.

The national parks and monuments are administered by the National Park Service in the Department of the Interior. Prior to 1916 they were patrolled by the Army, but in that year the Park Service was set up to "conserve the scenery and the natural and historic objects and the wildlife therein, and to provide for the enjoyment of the same in such manner and by such means as will leave them unimpaired for the enjoyment of future generations." The charge is a challenging one to fulfill, but one to which the people of the Park Service are remarkably dedicated. Their dedication spills over to us who visit the parks, imbuing in us a sense of reverence for the wonders themselves and reminding us that they are ours to enjoy and protect.

The National Park Service is faced with 40 million visitors a year. Some parks have become so crowded at their peak season that Senator Frank E. Moss of Utah remarked, "Many of our national parks are no longer a place of escape and repose but a massive traffic jam as nerve racking as a five o'clock rush." The problem might take care of itself if the number of travelers stayed the same, for those who visit a place one year go elsewhere the next, and places that are not so well known are attracting more visitors each year. But the number of visitors does increase. Many parks have loop roads that circle the main points of interest, with lodges, restaurants, and camp grounds at intervals. The traffic is almost entirely confined to these roads, and the crowds to the most spectacular sights. The Park Service is studying ways to relieve the pressure, perhaps by building roads into some of the lovely areas not now visited. Facilities for lodging and camping might have to be removed from the scenic centers so that people will be more spread out. The public will gain, for their visits will be a more pleasurable experience and embrace more varied country.

The Bureau of Land Management in the Department of the Interior conserves and manages 460 million acres of public domain in Alaska and eleven far western states. Their policy is that of multiple use, involving timber production, watershed protection, fish and wildlife development, livestock grazing by permit to ranchers, and recreation. This land embraces forested mountains, open plains, and desert. Some that you might not call scenic is important to wildlife which shares the area with cattle and sheep, and some is rich hunting ground for rock hounds and fossil collectors, and attractive to photographers and bird watchers. Much of it borders fishing waters and some offers hunting in season.

Many states have established recreational areas within their borders. The state parks cater to a great need for vacation spots within easy reach, where people can go for a weekend or a longer period without having to travel far or spend much money, or where they can just spend an afternoon. Some of the loveliest spots in each state are being preserved—some that rival the national parks on a small scale, and some that are of great historical, geological, biological, or archaelogical interest. Many of them border newly made reservoirs where water sports and fishing can be enjoyed by people who never before had the opportunity. Some states have developed interpretive natural history programs that attract and inform people of all ages.

The Forest Service in the United States Department of Agriculture has the management of 188 million acres of forest land organized into 152 national forests. The forests serve many purposes, among them watershed protection, supplying timber, and grazing for cattle. Some are set aside chiefly for recreation, with limited grazing, and with logging allowed only in areas away from the most scenic parts. The Forest

The Cloud Peak Wilderness Area in the Bighorn National Forest, Wyoming. (*U.S. Forest Service. Photo by Jay Higgins*)

Service has the mammoth job of maintaining their natural beauty while providing camp grounds and picnic areas, and places for fishing, swimming, boating, hiking, and winter sports. Roads, clean camp grounds, grills for cooking, parking places for trailers, sanitary facilities, firewood, and drinking water—all are meticulously maintained. Included in the national forests is a wealth of magnificent scenery.

Here and there, untouched by man's activities by great good fortune, are small primeval areas that are to be kept forever wild. No logging or livestock grazing is permitted. You may find one adjacent to a picnic area or just off a main road. You are welcome to wander through them to see what the forests were like when only the Indians roamed them. They are small samples of the climax forest that once covered the whole area, where the workings of nature have not been upset by the workings of man. Even the casual picnicker will sense the difference between their natural condition and that of forests that have been altered in some way.

On a larger scale are the wilderness areas, of 50,000 acres or more. Here there are no permanent residents, no roads, ski areas, lodges, or other man-made facilities. Each area is large enough so that you can hike, canoe, or ride horseback for a week or more without backtracking, and become completely separated from a routine existence and absorbed into nature. The Wilderness Act of 1964 gave statutory protection to areas already established by the Forest Service, and provided that 52.1 million acres could be added not only from Forest Service land but from other federal lands as well.

Some dissension was heard from people who desired roads and facilities and permission to use motor boats in the Wilderness Areas. But it was and is the aim to keep these lands unsullied by fumes and traffic, free from the sights and sounds of civilization, as pure and beautiful and wild as they were when we inherited them. If we let them slip away they can never be recovered. We need them so that future generations can know the peace and grandeur of an untouched world.

Index